Simple, Low Cost Electronics Projects

By Fred Blechman, K6UGT

LLH
Technology Publishing
Route 2, Box 99M
Eagle Rock, Virginia

ISBN 1-878707-46-9
Library of Congress catalog number: 97-77994

Printed in the United States of America.

Cover design: Sergio Villarreal, San Diego, CA
Production services: Greg Calvert, San Diego, CA
Original art and schematics: Raoul Patterson, San Diego, CA
Editorial development and project management: Vince Everett, Memphis, TN

Route 2, Box 99M
Eagle Rock, VA 24085

Table of Contents

Foreword

Today, we more and more live in a virtual world where pretend and fantasy are just a mouse click away. While little harm, and some good, may come from this, as an electronics instructor I get concerned when a student tells me he'd rather "build" circuits on a computer screen than in the laboratory or workbench at home. "Electrons don't flow on schematic drawings," I tell him, "they move in real devices, circuits, and systems."

To learn electronics, you must *do* electronics. You must construct circuits by bringing resistors, capacitors, transistors, ICs, and other electronic components together to create a working device—be it a siren, telephone line analyzer, voice-operated switch, digital alarm clock, function generator, or FM transmitter. You must heat up a soldering iron, insert components into a circuit board, and apply power. And when you do, when the device you built lights up, beeps, counts, or responds to your voice, you'll not only "know" electronics in a way no cyber-hobbyist ever could, you'll feel a pride of accomplishment unmatched in any fantasy world.

This book gives you a chance to enter the real world of electronics, and not just because it contains 22 easy to build, fun, and useful electronics projects. There are other project books that will give you that. No, the reason this is the book to begin your exploration of electronics is simple—it is written by a man who learned electronics the way you will be learning it—by doing it. Fred Blechman has immersed himself in electronics for over three decades. As far as I know, he has never taken a course in the subject. Yet, with over 700 articles and five books to his credit, Fred has much to offer the electronics enthusiast. Written in a clear, conversational style, the author never forgets what it's like to be a novice. In his latest book, Fred takes you with him as, together, you build, learn, and have fun with electronics. Who knows where that can lead. . . .

Ronald A. Reis
Electronics Instructor
Los Angeles Valley College

Foreword

Preface

I worked for years in the aerospace industry alongside degreed electrical engineers who had no real interest in electronics; it was just a job to them. They were "systems" designers who dealt only in block diagrams and flow charts, but couldn't design a circuit if their life depended on it!

They did not have electronics as a hobby. They did not know the resistor color code or the different types of capacitors. They knew formulas and theory, but couldn't reduce them to practice, and depended on circuit designers for those "mundane" tasks.

I developed a passion for electronics when I decided to build and fly my own radio-controlled (R/C) model airplanes. I had been a Navy fighter pilot, so the flying part was easy. But when two of my R/C models flew away while using my kit-built radio transmitter and receivers, I decided I better find out what was happening.

I read electronic magazines and started building circuits and test equipment from kits. Not only did my planes no longer fly away, but in order to fly them on less-used frequencies, I got my radio amateur "ham" license (K6UGT). This led me further into building and modifying electronic equipment, and I soon found myself writing articles for the very electronic and ham magazines from which I had started learning!

There is probably no better way to learn a subject than the "hands-on" approach. You can read about flying an airplane, for example, but until you're behind the controls you are not really learning how to fly.

Similarly, building electronic circuits from "scratch" or from kits teaches you about the various components and how they inter-relate in a circuit. You learn soldering and troubleshooting. You develop confidence and the ability to work with electronic equipment that can make you more valuable on your job. And these days, when electronics is so pervasive, that can mean just about ANY job!

My interest and knowledge made me the "copy machine guru" at my aerospace job; when anything went wrong, they called me. As I moved up into systems design and contracts, I had a better feeling with what was reasonable, electronically, than many of the all-theory-no-practice engineers.

In later years my lack of fear of electronics led me into microcomputers. I immediately began to write magazine articles and books using simple word-processing instead of fighting with a typewriter. If the computer went "down" I had enough confidence in my electronic ability to tackle basic repairs.

So it is with YOU in mind that I have assembled many simple projects in this book—a book intended to teach you some simple electronic fundamentals by having you assemble simple low-cost electronic projects. This book is for YOU if you have a real interest in electronics, but are a relative novice. You are a beginning or intermediate hobbyist or experimenter, and are far enough along to be familiar with components and common schematic symbols. However, you find most electronic books and magazine articles too full of formulas and too lacking in simple explanations.

My circuit descriptions are intended to be tutorial, and I apologize in advance if some seem too oversimplified. Those are for beginners. On the other hand, some may seem too complicated, in which case thinking them through will teach you something!

In any case, get out your soldering iron, find an appropriate workspace, and HAVE FUN while learning!

Fred Blechman

Fred Blechman is a former U.S. Navy F4U Corsair fighter pilot and has written over 700 magazine articles and six books about electronics, microcomputers, and flying since 1961.

CHAPTER 1

A Telephone Hold Button

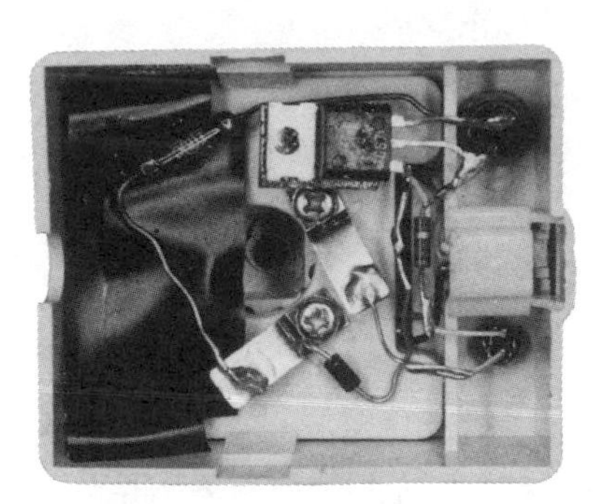

This telephone hold button will allow you to hang up the phone you're on and keep the phone line "alive" as you move to another phone. You won't have to go back and hang up the first phone. A printed circuit board layout is provided, but a PC board is not required to build this circuit.

"What? Only six parts—and not even a battery? Is this worth the space for a construction article in a book?"

That could well be your reaction when you look at the simple circuit for the telephone hold button, as shown in Figure 1-1. What isn't obvious is the number of things you can learn from building and using this circuit. Not only that, you'll probably want to build several of these "buttons," perhaps one for every phone in your home or office!

The telephone hold button allows you to put any incoming or outgoing call on "hold," with a red indicator showing the other party is waiting on the line. It works with all phones on that line; hang up one phone, and take the call on another. Or just put someone on hold while you leave the phone to get something or someone, and the telephone microphone is muted.

When you pick up the phone on that line, the indicator goes out and the phone operates normally. When the call is completed you won't have to run back to hang up the phone you answered or from which you called. Also, you won't have those times when you FORGET to go back and hang up the phone, leaving the phone line "busy" to callers—until you notice the phone off the hook!

If you're ingenious, you can build this circuit right into existing telephones. At the time this book was published, there was an inexpensive kit of parts available. We'll describe this kit and the required parts later in this chapter.

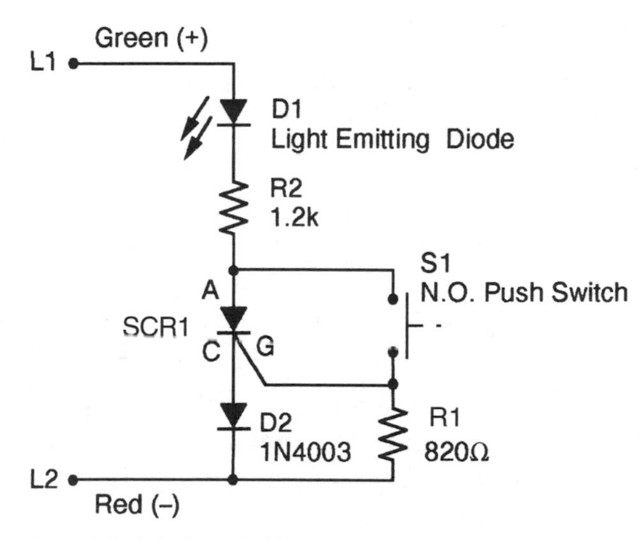

Figure 1-1: Telephone hold button schematic.

Inside Your Telephone

Telephones are everywhere, and pretty much taken for granted. You pick up the handset and you get a dial tone. You dial a number (usually by pushing buttons) and the other party answers. When you're finished, you hang up. You probably never think of the amazing technology behind it all, especially when calling long-distance or internationally.

What started as simply two wires and a battery between two instruments is now an almost incomprehensible conglomeration of wires, relays, amplifiers, undersea cables, fiber-optics, microwave towers, geostationary satellites, and "black boxes" of all sorts—always changing, improving, and getting more complex.

But some things remain the same. Your standard phone has at least two wires coming into it, usually red and green. (In multi-phone installations, you have various solid-colored and striped wires.) The green wire carries a positive voltage in relation to the red wire. When the phone is not in use ("on hook"), this voltage is about 45 volts direct current (DC), and the current is essentially zero (actually, around 20 microamperes, with the actual value depending on a number of things).

When the receiver is raised "off hook," this voltage drops to about 5 volts DC, and a significant current (about 25 milliamperes, with the actual value depending on several factors) flows through the phone lines. This milliampere current flow signals the central station that your phone is off-hook, and the dial tone is connected to your line. As long as this approximate amount of current flows, the line stays connected.

If you want to keep the line connected ("hold") when you hang up the receiver, you must find some way to keep the current flowing, thus fooling the central station into thinking you're still connected. The telephone hold button does just that.

Component Descriptions

Although this is a simple circuit, some of the components require further explanation. This will help you to understand how the telephone hold button works. Later on, I'll tell you how to substitute parts for those specified if you decide to "roll your own" instead of getting the parts kit offered.

The light-emitting diode (LED; see Figure 1-2) has become quite common, replacing the pervasive neon and power-hungry incandescent pilot lights and indicators used years ago. The LED is merely a one-way electronic gate that glows (red is the most common color) when the "anode" has a positive "breakdown" voltage with respect to the "cathode." This is usually a small voltage, as low as 1 volt to start emitting light, so very little power is used. The more current, the brighter the glow.

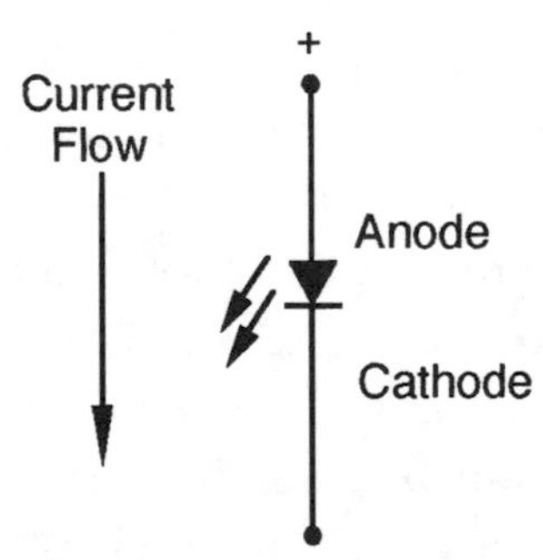

Figure 1-2: Schematic symbol for a light-emitting diode (LED)

However, an LED used by itself will "avalanche." That is, once the breakdown voltage is reached, current will increase and destroy the LED anode/cathode junction unless the current is limited. A resistor is used to limit current, and the resistor also limits LED brightness.

LEDs can fail. Too much current (from either too much forward voltage, or too low a value of resistance, or both) can burn out an LED. Also, if you apply too much voltage in reverse (positive on cathode with respect to anode), you can burn out the junction.

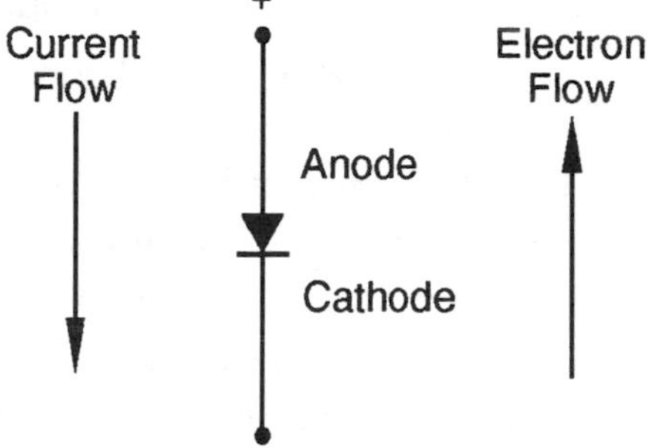

Figure 1-3: Schematic symbol for a semiconductor diode

The diode (also called "rectifier") used in this circuit (Figure 1-3) is a one-way electronic gate that does not glow. It allows current to flow in only one direction. When the voltage on the anode is approximately 0.7 volts higher than the cathode, current flows. (Remember, we're talking current here, not electrons. Current flows from positive to negative—or less positive—and electrons flow in the reverse direction, attracted to positive.) This results in a small voltage drop of about 0.7 volts across the diode junction.

The SCR (silicon-controlled rectifier), as shown in Figure 1-4, has three terminals: anode, cathode, and gate. In order to conduct current, the anode must be positive with respect to the cathode, like any rectifier (diode). However, unlike a regular rectifier, the SCR needs to be "triggered" by a minimum positive voltage at the gate terminal. In effect, positive voltage provides a "latching current" at the gate that unlocks the anode/cathode path, allowing essentially free current flow. Like any diode, there must be a "load" to prevent too much current from flowing, burning out the junction. But there's another feature of an SCR that makes it especially useful: once the trigger voltage is removed, the SCR continues to conduct from anode to cathode unless it is "starved" below a "holding current" threshold. I'll talk about this again later in this chapter.

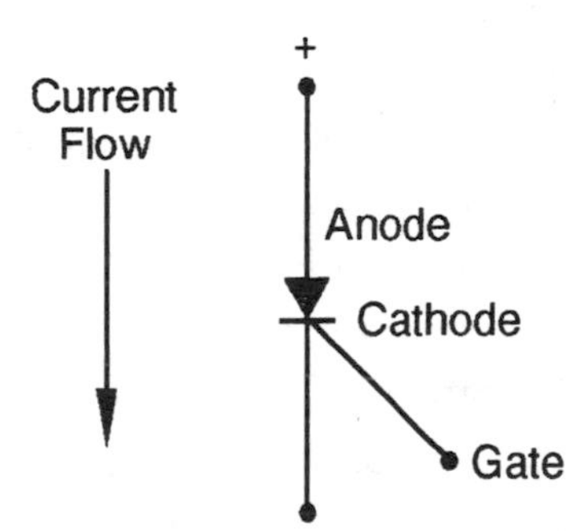

Figure 1-4: Schematic symbol for a silicon-controlled rectifier

Circuit Description

Looking at Figure 1-1, you'll see that D1, R2, SCR1, and D2 are all connected in series across the phone line, with the GREEN wire assumed positive. If the phone is on-hook, other than a small "leakage" current, nothing is happening. Why? The SCR has no positive voltage on the gate, so the gate is closed, even though the voltage from GREEN to RED is about 45 volts DC.

When you pick up the phone, it is off-hook and the GREEN to RED voltage is about 5 volts DC. It doesn't matter if you're making a call or receiving one. To put the other party on hold, press the S1 switch. Suddenly the top of R1 forms a voltage divider with R2. The positive voltage at the top of R1 (a ratio of R1/(R1+R2), or 820/2000) triggers SCR1 into conduction, permitting current to flow through D1, R2, SCR1, and D2. The LED glows slightly. Now when you hang up the phone, the voltage increases to almost 30 volts and the LED glows brightly. Release S1. The LED stays bright and the line is held because sufficient current is flowing through SCR1.

But that's not the end of the circuit action. When any phone on that line is picked up (even an extension) the line voltage drops to 5 volts again, and the LED should turn OFF. At this low voltage, the SCR becomes current-starved and stops conducting. That's exactly what you want. The hold button circuitry is now again dormant, and, when you hang up, the line is released.

Diode D2 is used here to protect the circuit from reverse voltage when connected to the phone lines, especially since the GREEN and RED voltages are sometimes reversed in older phone installations. (The old rotary-dial phones did not care about voltage polarity. Many tone-type phones will not work if the phone line polarity is reversed.)

Parts Substitution

As is common with simple circuits, you can substitute components in place of the ones specified in the parts list. My intention here is to specify parts commonly available from electronic suppliers.

For example, you can use resistors of smaller (1/8 watt) or larger (1 watt or more) power ratings, and they can be made from substances other than carbon. There is nothing very critical about the resistance values, so you don't need to look for expensive 1% tolerance or even 5% tolerance resistors. In fact, as I'll mention later, one of the resistor values may need to be changed to work with your particular telephone line!

LEDs come in various colors, sizes, and shapes. There is no need for any special LED here, but I suggest you get the common red "jumbo" type (cylindrical, about 3/16-inch diameter with a dome top), since they are the brightest for the lowest cost.

The commonly available 1N4003 diode specified in the parts list is a power diode, intended to carry considerable current, and withstand significant reverse voltage. It is rated to carry one ampere (1 A) of forward current, and withstand 200 volts of peak inverse voltage (PIV). Since this is connected essentially across the phone line, and voltage spikes can occur, it's wise not to use a lower-rated reverse voltage diode. However, the current capacity need not be more than 50 milliamperes (mA).

There are many types of SCRs, some capable of carrying very large currents. In this circuit, the specified C106B1 is a common type rated for 6 A and 200 PIV. Like the diode just described, you only need about a 50-mA current capacity in this circuit, but stick with the 200 PIV. The Radio Shack catalog number 276-1067 SCR (priced at 99¢ at the time this book was being written) looks like it should be perfect, but read the CAUTION further on. I recommend the Mouser 519-T106B1 sensitive SCR (currently priced at 79¢).

Figure 1-5 shows the layout for a printed circuit board for this circuit. Follow this pattern if you want to etch and drill your own board. If you buy the parts kit, an etched and drilled board will be included.

The S1 switch can be any sort of normally-open pushbutton type that closes when pressed, and opens when released. The most popular type is the kind with a red button on top, as shown in the Figure 1-6 pictorial. If you are going to install the telephone hold button circuit inside an existing telephone, you may need to find a miniature pushbutton switch to fit the space available.

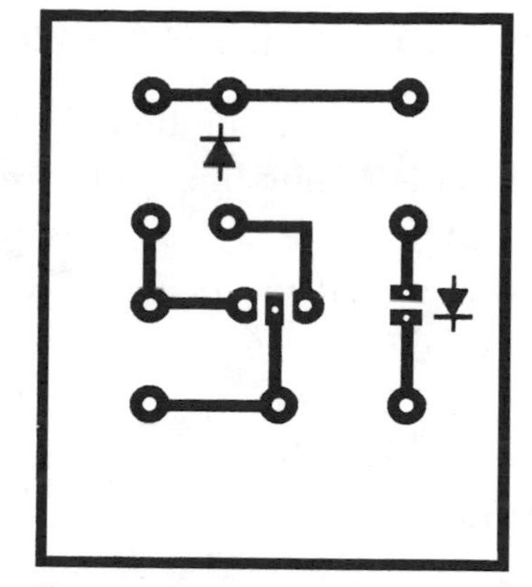

Figure 1-5: Printed circuit board layout pattern; board should be exact size as this figure

Construction

If you build the telephone hold button kit available from Centerpointe (see parts list), it will take about 15 minutes to build. Figure 1-6 shows how the parts are mounted. (Note that the documentation that comes with the kit may transpose L1 and L2 locations. Use L1 and L2 as shown in Figure 1-6.)

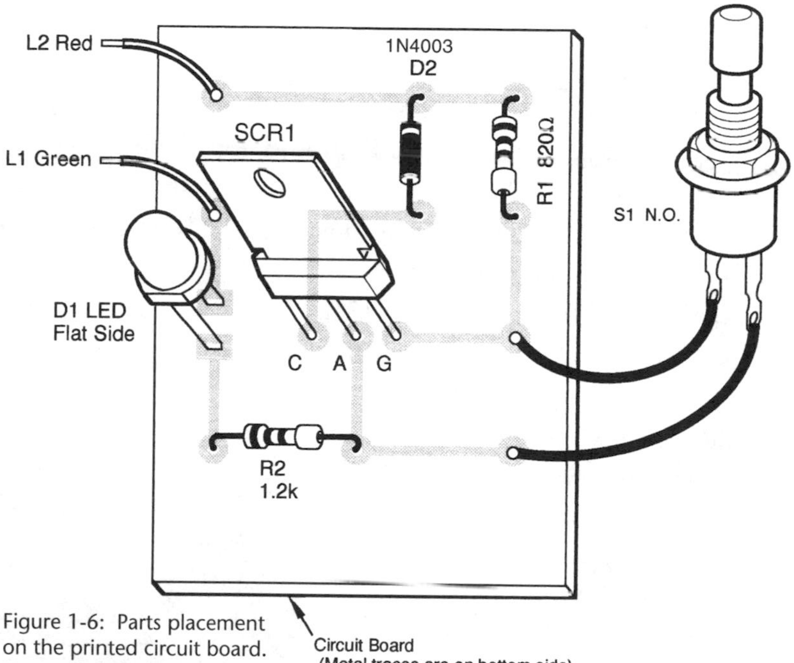

You do not need to use a printed circuit board if you are making this project from "scratch" with parts you have or obtain elsewhere. Point-to-point wiring using a perforated board will work fine. For example, Radio Shack sells three sizes of phenolic boards with 0.042-inch-diameter holes spaced in a standard 0.100- by 0.100-inch grid pattern, ideal for home-brew circuits and small projects. The Radio Shack catalog numbers at the time this book was published were 276-1395, 276-1396 and 276-1397. I suggest getting the largest one (6 inches by 8 inches) since you can cut it into many smaller boards with a fine saw. You can obtain similar boards from other electronics parts retailers.

Figure 1-6: Parts placement on the printed circuit board. Note carefully the orientation of some parts!

Circuit Board
(Metal traces are on bottom side)

Simple as this project is, you still need to be careful about several things. The LED must have its flat side (cathode) placed as shown, D2 must have the colored band (cathode) placed as shown, and SCR1 should be facing as shown, so the cathode and anode leads are not reversed. Leads L1 and L2 go to your telephone, as explained later.

If you are intending to install the hold button circuitry inside a telephone, the printed circuit board (1.6 by 2.0 inches) will probably be too big; you'll have to make your own design or use perforated board. If you use your own parts, read the CAUTION further on.

Testing and Troubleshooting

The best way to test the finished unit is to use clip leads to temporarily connect it to the phone line, just as you intend to install it. This way you'll have access to both the high and low phone line voltages.

First, you'll need to open up the base of your phone and identify the two wires that come into the base and power your phone. These are normally GREEN (positive) and RED, but can be different colors on multi-line phones. If in doubt, use a voltmeter and look for about 45 volts DC with the phone not in use.

Connect the L1 (GREEN) lead of the hold button to the positive voltage terminal of your phone line, and L2 (RED) to the other terminal in your phone. Nothing should happen.

Pick up the phone, listen for the dial tone, then dial your own number. You'll hear a busy signal. Press S1 and the LED will probably glow dimly. Continue to press S1 and hang up the phone. The LED should get much brighter. Release S1 and the LED should stay on and glow brightly.

After a slight delay to be sure the LED stays on, pick up the phone again. You should still hear the busy signal (NOT the dial tone!), and the LED should go completely out.

If the hold button isn't working properly, check the orientation of D1, D2 and SCR1, as described earlier. Make sure all solder connections are good, and no solder "bridges" are shorting connections. If your unit still doesn't work (especially if it's a "homebrew" job), read the CAUTION that follows later.

Installation

As noted earlier, you'll probably find the printed circuit board too large to put inside a regular telephone if you build the kit in the parts list. In such a case, you can place the telephone hold button board alongside your telephone or attach it to your phone with double-sided tape. You can be as neat or messy as you please. Just make sure you can easily press the switch and see the LED.

Figure 1-7: Here is a typical way to install the telephone hold button on the side of a standard desk telephone.

Open up your telephone base and connect the hold button GREEN and RED wires as just described for "Testing." The hold button should work as described. Figure 1-7 shows a typical installation. It may not look too elegant, but it gets the job done!

If you like a challenge and are handy, you can wire the parts inside your telephone base or inside an adapter that your phone can plug into. Make sure the switch and LED are mounted to be convenient and visible. Figure 1-8 shows this circuit built inside a four-prong modular telephone adapter. The button and LED are installed outside on the top of the adapter.

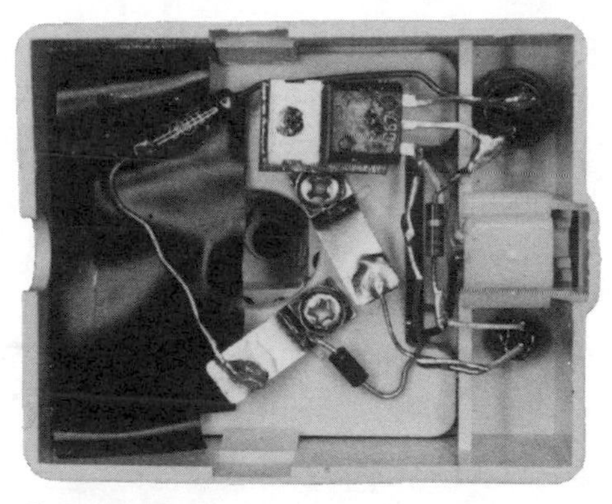

Figure 1-8: The entire circuitry for the telephone hold button can be installed inside a four-prong modular adapter, as shown here.

CAUTION! Beware the SCR Trap!

The SCR is the "heart" of this circuit, and not all SCRs will work the same. Let me tell you what happened to me.

I built three telephone hold buttons, since there are three phone devices in this house that I frequently pick up and have to put down to go to a different telephone in another room. I built the first hold button from the kit described in the parts list. It worked like a charm! I decided to build two others from parts I had around. That's when the frustrations started.

After a quick search through my various parts boxes, it seemed I had all but the 820 Ω resistors (what an odd value!) and the SCRs. Since I was hot-to-trot and a Radio Shack store was nearby, I went there. They do not carry 820 Ω resistors as regular stock items (special order only), but I got two of their catalog number 276-1067 SCRs (identical ratings and similar in appearance to the SCR furnished in the parts kit).

Scrounging into some old parts boxes I hadn't used in over 20 years, I found two 820 Ω 1/2-watt resistors. Using a 1 inch by 1.5 inch perforated board, I connected up the parts with point-to-point wiring. It didn't work. I carefully checked the wiring and found I had two SCR leads reversed. My fault, but very easy to do with point-to-point wiring—it seems the simpler the circuit, the less careful you are likely to be.

But it still refused to work when connected to the phone. It acted completely dead. Hmmmm. Using my trusty multimeter, I found the GREEN and RED wires on this particular phone line (used with a rotary phone) had the polarity reversed! So I reversed the connection to the hold button circuit—and it seemed to work. Then I noticed it refused to reset when the phone was put back on the hook; the LED stayed ON instead of going OFF. Apparently the off-hook voltage was too high to current-starve the SCR, and it kept conducting.

The solution was lowering the value of R1 from 820 Ω to 680 or 470 Ω. This has the effect of stealing some holding current from the SCR (thus cutting off the SCR at a higher off-hook voltage), but still allowing it to trigger when S1 is closed. Since the Radio Shack SCR was not the specified C106B1 intended for this circuit, it probably has slightly different characteristics. Okay, two down and one to go.

Number three was even more trouble. Built just like the homebrew unit #2, it refused to work properly. The LED would go on, but not stay on. I tried everything I could think of. I checked the wiring thoroughly—it was okay. I changed R2 to a lower value. I checked the phone line polarity.

Finally, I looked around and found a C106B1 SCR in another as-yet-unbuilt parts kit, and replaced the Radio Shack SCR with the SCR. The hold button now worked perfectly! Apparently the Radio Shack SCR required too much holding current. Once again, I suggest a better choice might be the Mouser 519-T106B1 sensitive SCR.

Summary

The telephone hold button is such a handy and inexpensive device that you might want to make one for each of the phones in your home or office. It will save you from going back to hang up a phone when you go to another extension, or—worse—forgetting to go back and leaving the phone line "busy."

But be careful about making homebrew versions with junkbox parts!

PARTS LIST

D1:	light-emitting diode (see text)
D2:	1N4003 silicon rectifier (see text)
R1:	820 Ω 1/8 W carbon resistor (see text)
R2:	1.2 kilohm 1/8 W carbon resistor
S1:	SPST normally-open pushbutton switch
SCR1:	C106B1 silicon-controlled rectifier (see text)

NOTE: *At the time this book was being written, a complete kit of parts, including an etched and drilled printed circuit board, was available from Centerpointe Electronics, Inc., 5241 Lincoln Avenue, Unit A7, Cypress, CA 90360; toll-free order line was 800-272-2737. Their stock number was 9992, and the cost per kit was approximately $5.00. Please remember that availability and pricing may have changed since publication of this book.*

CHAPTER 2

A Telephone Remote Ringer

This telephone remote ringer lets you know someone is calling when you are at a location where you don't have an extension phone connected. It uses only five parts, and a common 9-volt battery lasts for about 36,000 "rings." No printed circuit board is needed.

It's really convenient to have extension telephones on your patio, in your workshop, in the garage, or in any number of other places. I have eight telephones connected to one line in my five-bedroom two-story home—and I still find myself running to answer the phone from someplace where I don't have an extension! Since phones and phone wire are so inexpensive today, it is not too difficult to end up with a lot of extensions on a single phone line. And modular jacks and adapters make it relatively easy to extend your phone lines all over the place.

But there may be some problems. For one thing, if you have too many phones off the hook at the same time, you will seriously reduce the volume on each. But more of a problem is whether a particular phone will ring when a call is coming in!

Why? Because the phone company will supply only a limited ring current, typically enough to ring five standard telephones. Each telephone has a REN (*r*inger *e*quivalence *n*umber) that is supposed to be listed on each phone's nameplate. If all the RENs add up to over 5.0 on a single phone line, the phone company does not guarantee all the phones will ring.

If you have a phone in the kitchen, one each in a couple of bedrooms, one on the patio, one in the garage or workshop, and a remote or two, all connected to the same phone line, chances are you're exceeding a REN of 5.0 and some of your phones will refuse to ring.

What you need is a "telephone remote ringer" wherever you want to know your phone is ringing. This can be where you have an extension phone that doesn't ring, or just someplace where you don't even have a telephone but you want to know when someone is calling.

The telephone remote ringer described here is extremely simple, and uses only five parts and a small battery. It takes very little phone power to operate, and a standard small 9-volt battery should be good for about

36,000 rings! The installation is as simple as plugging it into your phone line and forgetting it.

Circuit Description

Figure 2-1 shows the simple schematic. Capacitor C1 blocks the approximate 48 volts direct current on the phone line when all phones are hung up. No current flows from the phone line. When the ringing signal—about 90 volts AC superimposed on the 48 volts DC—comes in from the phone company, the AC signal passes through C1 and through voltage dropping resistor R1. Diode D1 effectively clips (shorts out) the negative pulses, so only positive half-wave voltage appears at pin 1 of OPT, a 4N26 optoisolator.

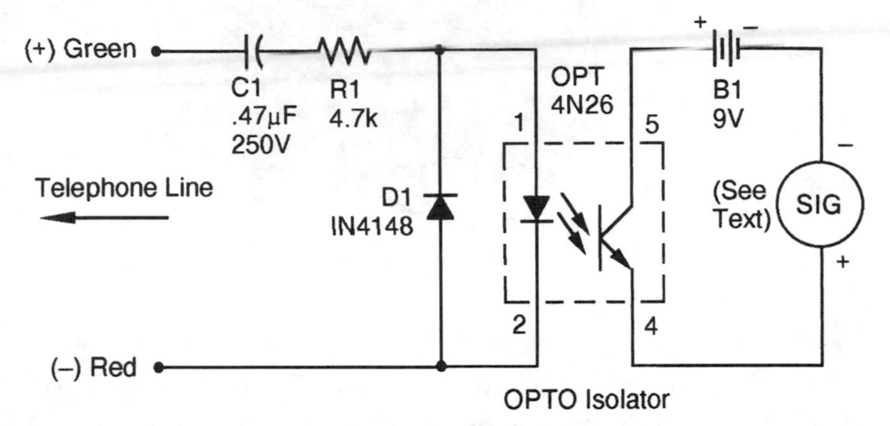

Figure 2-1: Schematic diagram for the telephone remote ringer.

The optoisolator physically looks like a six-pin integrated circuit with dual in-line pins. All it has inside, however, is an LED (light-emitting diode) and a phototransistor, with the case eliminating all outside light. Instead of using a base current to control conduction, the phototransistor is enabled by light. With no light, no current can flow from the collector (pin 5) to the emitter (pin 4). However, when the internal LED lights, the phototransistor conducts, allowing current to flow from the positive battery terminal, through the phototransistor, through the SIG signaling device, and back to the negative side of the battery. Pretty simple!

You may need to do some searching to find the 4N26, as it is not a common device. At the time this book was being written, Mouser Electronics sold the 4N26 as stock # 512-4N26 at a list price of 38¢ each. (Of course, price and availability may have changed since this book was published.)

An advantage of the optoisolator, in addition to its simplicity and low driving current, is that the output circuit—the battery and signaling device—is totally isolated from the telephone input circuit. Also, external light has no effect on its operation.

The signaling device can be a Mallory Sonalert, or any of a variety of buzzers or piezo-electric sounders. To keep battery life long, I used an SC628 Sonalert that I happened to have in my junk box. When the phone rings, it only uses about 0.5 milliamperes and puts out a fairly loud and recognizable signal. From battery life charts I have, it appears a standard zinc-carbon 2U6 9-volt battery should last for 50 "ringing" hours, or about 36,000 rings at 12 rings a minute!

Unfortunately, the original Sonalerts cost $15 to $30 these days. Mallory now makes a Sonalert II (available at the time this book was written from

Mouser as stock # 539-MSR320 for $3.15). An alternative available when this book was written was the piezo sounder from Radio Shack (catalog # 273-060, $3.29). Avoid cheap buzzers, since they often use too much current and would greatly shorten battery life.

Oddly enough, the toughest thing to find will probably be the 0.47 µF 250 WV capacitor. If you are an old timer with a junk box full of old, big, wax-covered radio capacitors, no problem. But with the low voltages used in most equipment today, high capacity, high voltage, non-polarized capacitors are hard to find. One source at the time this book was written was Mouser; their stock #1430-2474 was priced at 52¢ each.

Construction

This circuit is easily constructed on a perforated board using point-to-point wiring. You'll need a small plastic or metal cabinet large enough to hold the battery and signaling device. Be sure that D1 and the signaling device are wired properly with regard to polarity.

No special layout precautions are required, except to keep the telephone wires from touching the case if you use a metal cabinet. I used a small surplus metal cabinet and covered it with walnut-simulated self-stick paper. A two-wire cable extends out to where it connects to the phone line. Figure 2-2 shows my completed unit.

Figure 2-2: Almost any small cabinet, like the one shown here, can be used to house the telephone remote ringer.

Installation

The telephone remote ringer must be installed in parallel with the phone line. Typically, this means just plugging it into a telephone wall jack with a standard RJ-11 modular phone plug at the end of a cable from your ringer. Or, you could install an RJ-11 phone jack in the ringer cabinet and use a standard phone cord (modular plug at each end) to connect it to the wall jack.

If you wish to use the telephone remote ringer where a ringer-disabled extension telephone is in use, you'll need to use a common RJ-11 1-to-2-jack adapter that converts one modular jack to two jacks. Radio Shack and Mouser are two sources of the modular plugs, jacks and adapters you may need.

The important thing is that you pay close attention to the polarity when you connect the ringer to the phone line. The GREEN wire shown on the schematic must be connected to the positive side of the phone line! A voltmeter (about 48 volts DC with all phones hung up) can be used to identify the positive phone line wire.

Troubleshooting

Once connected to the phone line, call this line from another telephone, or ask a friend to call you. If the ringer doesn't sound, test the battery and signaling device. You do this by simply connecting a clip lead from pin 5 to pin 4 of the 4N26 optoisolator, simulating phototransistor conduction. If the battery and signaling device are good, you should hear the sound.

If this works, then you can test the LED portion of the 4N26 by placing a small positive DC voltage (say, a 9-volt battery with a 1000 Ω resistor in series) to pin 1 of the 4N26, with the negative side to pin 2. The signaling device should sound.

The ringer still doesn't work? You could have the GREEN and RED wires reversed at the phone line, giving you the wrong polarity. There is also the possibility of an open C1 capacitor or R1 resistor, or a shorted D1 diode—or a bad solder connection. Also, D1 and SIG are polarity sensitive.

This is a very simple project, but it results in a very useful device. I've used several since 1978!

PARTS LIST	
B1:	2U6 9-volt transistor radio battery
C1:	0.47-µF 250 WV capacitor (see text)
D1:	1N4148 or 1N914 silicon signal diode
OPT:	4N26 optoisolator (see text)
R1:	4.7-kilohm 1/4 W 10% carbon resistor
SIG :	Sonalert or other signaling device (see text)

CHAPTER 3

A Telephone Recording Beeper

This telephone recording beeper lets you alert all parties to a telephone conversation that it is being recorded. It can provide over 150 hours of service from a common 9-volt battery. No printed circuit board is needed to build this circuit.

Penal Code Section 632 of California law specifies that recording a telephone conversation without permission of all parties involved (with some exceptions) is illegal, and punishable by a fine of up to $2500 or one year in jail, or both. Most other states have similar statutes.

Section 632 came up in the O.J. Simpson murder case preliminary hearing. It was disclosed that defense attorneys had recorded a telephone conversation with limousine driver Allan Park without Park's notification or permission. Information obtained during this phone conversation was then introduced in court during the defense's cross examination of Park—until the prosecuting district attorney pointed out that this recorded information had been obtained without the witness's knowledge or consent!

California law specifies that calls cannot be recorded unless (1) everyone on the conversation agrees, or (2) a beep-tone warning can be heard every 15 seconds. Most states have a similar requirement.

The telephone recording beeper we'll discuss in this chapter puts a beep tone on the telephone line approximately every 15 seconds when it is turned on. It connects directly to your phone line, and uses an internal battery that lasts for 150–300 hours of use. A red light-emitting diode (LED) flashes with every beep to remind you the beeper is turned on.

Circuit Description

Figure 3-1 shows the schematic of the telephone recording beeper, as well as the timing diagrams at three points in the circuit. A standard 2U6-type 9-volt transistor radio battery (carbon-zinc or alkaline) is used to power the device. When switch SW1 is closed, battery power is applied to integrated circuits U1 and U2.

The 555 timer, U1, is wired as an astable multivibrator, with resistors R1 and R2 and capacitor C1, controlling the charge-discharge timing cycle. Output pin 3 starts high and goes low at the end of each timing cycle. The initial cycle takes longer than subsequent cycles. At the end of each cycle, when U1 pin 3 goes low for about 0.2 seconds, section A of U2 inverts this to a high, thus lighting the LED, whose current is limited by resistor R3.

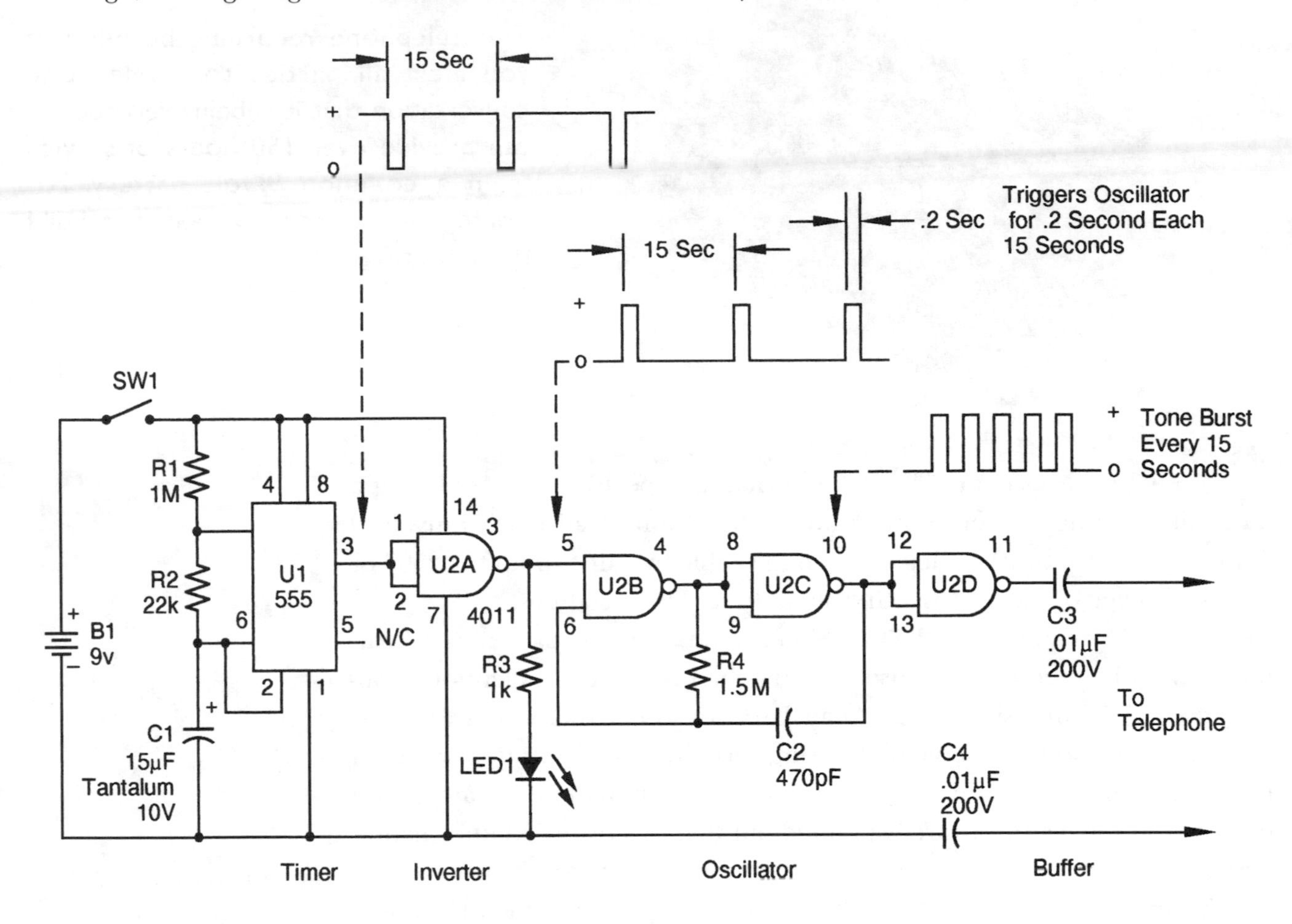

Figure 3-1: Schematic diagram for the telephone recording beeper.

This high is also applied to pin 5 of U2, which is the input to a gated oscillator composed of U2 sections B and C, resistor R4, and capacitor C2. The oscillator frequency (established by the values of R4 and C2) passes through buffer U2 section D and DC blocking capacitors C3 and C4, which connect directly across the phone line.

If you were to attach a properly adjusted oscilloscope to the circuit points indicated on Figure 3-1, you should see patterns similar to those shown.

The switch assures that there is no battery drain when the switch is open. Once the switch is closed, the battery drain averages about 3 milliamperes. A standard 2U6 carbon-zinc 9-volt battery should last for 150 hours of "on" time, and an alkaline battery about 300 hours. When not in use, of course, the switch should be turned off. The LED keeps blinking when the switch is on to remind you to turn off the switch.

Construction

Most components used in the beeper are readily available. You might have some trouble locating C1, the 15-µF tantalum capacitor. At the time this book was written, Mouser sold one with a 10-volt rating as their stock # 581-15K10V for 45¢ each. You should not use a regular electrolytic capacitor here, since their values vary considerably from the marked value and they degrade over time, which will seriously affect the timing.

High voltage, large capacity, non-polarized capacitors were common in the old AC radio days, but are now getting hard to find. C3 and C4 connect directly to the phone line, and need to be rated at a high enough DC working voltage (WVDC) to withstand the phone ringing AC voltage. Also, the larger the capacity, the easier the "beep" passes through the phone line. One source at the time this book was written was Mouser, who sold their stock # 140-PF2D103K 0.01µF 200 WVDC polyester film capacitors for 17¢ each.

The easiest means of construction is to use a perforated board with point-to-point wiring. It's probably a good idea to use sockets to mount the two integrated circuits, U1 and U2. Make sure capacitor C1 has its positive lead connected as shown in the schematic, and that the integrated circuits are not reversed with regard to their pin locations. (Looking from above, pin 1 is the upper left pin closest to the notch or indent at one end of an integrated circuit, and the pins are numbered counter-clockwise from pin 1.) The LED cathode (usually a flat spot on the base) should be connected to the battery negative side of the circuit, as shown.

Just about any small plastic or metal cabinet can be used, but make sure you have room for the battery, that the LED can be easily seen, and that the switch can be mounted for convenient access. Figure 3-2 shows how I fitted everything into a typical cabinet.

Figure 3-2: All of the telephone recording beeper's parts, including the battery, can fit inside a small plastic cabinet as shown here.

Installation

The connection to the phone line (standard red/green phone connection) can be with clip leads, or using an RJ-11 modular plug at the end of a cable. Alternately, you could use an RJ-11 modular jack mounted in your cabinet, and connect to the phone line with a regular line cord. Fortunately, line cord polarity is NOT a consideration with this design, since capacitors C3 and C4 provide total DC isolation while passing the beeper tone.

When the beeper is connected to the phone line, turn it on, pick up the handset, dial any number (to kill the dial tone), and listen carefully. Whenever the beeper LED flashes, you should hear a subdued tone for about 0.2 seconds. Don't expect a loud tone; this is not supposed to interfere with regular conversation, but simply to notify those using the phone that the conversation is being recorded.

Troubleshooting

What if you don't hear the "beep" about every 15 seconds (longer for the first cycle) when connected to the phone line and the phone is off-hook? You'll need a small audio amplifier to verify operation since the output of the beeper is not strong enough to drive a speaker or even an earphone (unless you have a good earphone, quiet surroundings, and great hearing!). Just connect the amplifier to the "phone" end of capacitors C3 and C4 and listen for the beep when the unit is turned on and the LED lights.

If the LED lights "on schedule" but you hear no tone, then U2 and its components are at fault. If the LED doesn't light at all, then U1 and its components are suspect.

Of course, be sure the battery is good and connected with the proper polarity, that C1 and LED1 are not reversed in polarity, and that pin 1 of U1 and U2 are in the proper location.

Finally, please be sure to determine what laws in your state and/or city govern the use of a beeper like this and other legal requirements placed on recording telephone conversations. A brief conversation with your lawyer about local laws could be very good insurance against getting into trouble from use of this beeper!

PARTS LIST

B1:	2U6-type 9-volt transistor radio battery
C1:	15µF 10-volt tantalum capacitor (see text)
C2:	470pF 50-volt ceramic disc capacitor
C3, C4 :	0.01µF 200 WV metal-film capacitor (see text)
U1:	555 8-pin DIP single timer
U2:	CMOS 4011 14-pin quad 2-input NAND gate
LED1:	Red light-emitting diode
R1:	1 megohm 1/4 W 10% carbon resistor
R2:	22 kilohm 1/4 W 10% carbon resistor
R3:	1 kilohm 1/4 W 10% carbon resistor
R4:	1.5 megohm 1/4 W 10% carbon resistor
SW1:	SPST slide or toggle switch
Optional:	14-pin and 8-pin IC sockets; RJ-11 modular phone plug or jack.

CHAPTER 4

A Telephone FM Transmitter

This circuit broadcasts both sides of a telephone conversation to a nearby standard FM receiver. It is intended for group listening, and can be built from scratch or from a kit that includes a printed circuit board and telephone cable with modular connectors.

Telephones are among the least secure communication devices, since it is so easy to "bug" a telephone line for surreptitious listening. The most common method is to simply connect somewhere across the phone line with blocking capacitors, and amplify the phone conversation into earphones. This direct connection, properly done, does not load down the phone line or affect the phone ringing, so it is normally not detectable. This is also illegal without a court order, unless the parties to the conversation are aware of the listening, or a 15-second "beep" is put on the line.

But what if you WANT the conversation to be easily heard by a group of people, such as a family call? The whole family can join in the conversation when you have a device that transmits both sides of the telephone conversation to a nearby radio. That's exactly what this easy-to-build circuit does!

Features

This telephone FM transmitter offers a number of handy features. For one thing, it does not require any battery, since it is powered by the telephone line whenever the telephone is "off-hook" (in use).

It transmits the phone conversation within the standard FM broadcast band (88–108 MHz), so any inexpensive FM radio can be used as the conversation monitor. A switch and LED (light-emitting diode) built onto the circuit board accommodate and verify proper connection to the telephone lines. Installation is a simple "plug-in" operation requiring no modifications to your telephone or the telephone outlet jacks.

At the time this book was being written, a $12.95 parts kit was available that included a silk-screened, drilled, and etched printed circuit board, all the necessary parts, and a standard modular telephone plug, jack, and cable. Ordering information is given at the end of this chapter.

Circuit Description

The schematic is shown in Figure 4-1. Although it looks at first like the telephone FM transmitter circuitry is connected ACROSS the telephone line, it is NOT! Actually, the circuit is only "alive" when the telephone is "off hook." This is when the handset is lifted so an internal normally-open "hook switch" closes and allows current to flow through the phone lines. This actually places the telephone FM transmitter circuitry in SERIES with the telephone line, as explained shortly.

Although not generally known, in a properly wired telephone line, the green wire is "earth ground" and the red wire is "minus" voltage. The voltage between them is about 45 volts DC when the phone is "on-hook" (hung up), with the green wire POSITIVE with respect to the red wire. When the phone is "off-hook" (in use) the voltage is about 5 volts DC, still with the green wire positive with respect to the red wire.

Refer to Figure 4-1 for the following discussion. When the telephone handset is lifted off-hook, telephone line current flows into telephone FM transmitter input IN1, out OUT1, through the now-closed telephone hook switch, into OUT2, through the telephone FM transmitter circuitry, and then returns to the telephone lines through IN2. The entire telephone FM transmitter circuitry is actually in series with the telephone line.

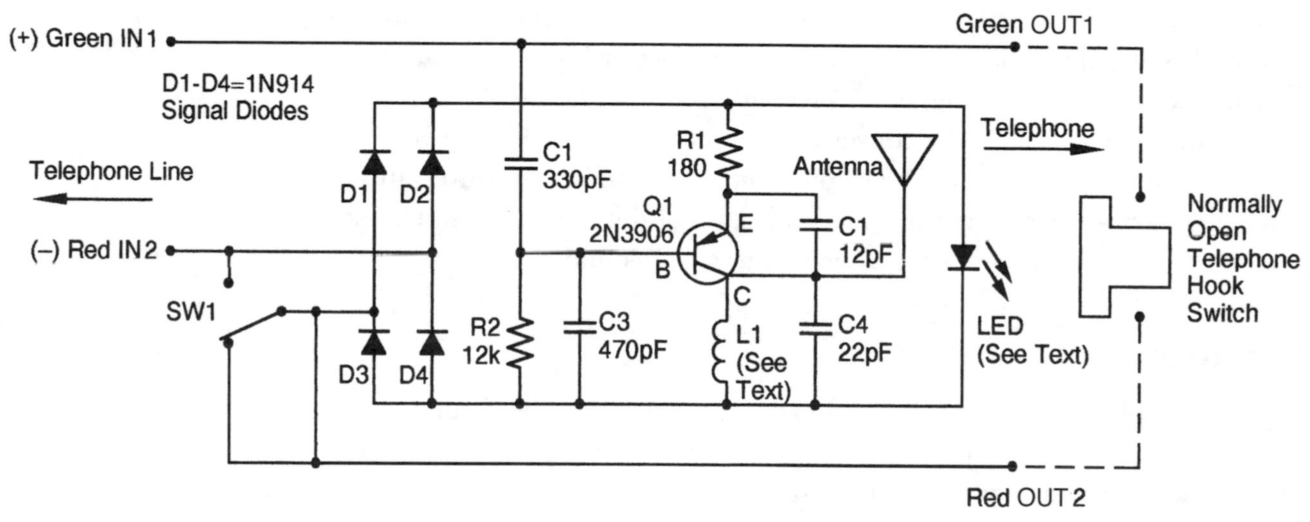

Figure 4-1: Schematic for the telephone FM transmitter.

Admittedly, this is confusing, since OUT1, OUT2, and IN2 are named only for their physical location between the phone line and the telephone rather than for their current flow direction. And the fact that the telephone company central office power is negative with respect to earth ground adds more confusion.

Following the schematic with switch SW1 in the position shown, assume the "green" telephone line is positive (with respect to "red") at IN1, and the phone is off-hook. Current flows through the telephone hook switch, through switch SW1, through diode D1, through the LED (which lights), then through diode D4, and finally returns back to the telephone line through IN2. The positive voltage at the anode of D1 powers the rest of the circuitry.

If, for whatever reason, the voltage at IN1 is NEGATIVE (with respect to IN2), the LED will not light. Putting SW1 in the OTHER position will bring diodes D2 and D3 into action, light the LED, and power the circuitry.

Either way, positive voltage appears at the emitter of PNP transistor Q1. At the collector, the parallel combination of inductor L1 and capacitor C4 form a tuned circuit somewhere in the FM broadcast band. Capacitor C2 provides feedback from collector to emitter to maintain oscillation. This oscillation is transmitted through the antenna as a "carrier wave" that can be picked up on a nearby receiver tuned to the oscillator frequency. Although not obvious, the oscillator frequency is controlled by the instantaneous voltage on the base of Q1, making this a voltage-controlled oscillator (VCO). If the base voltage were constant, the frequency would be unchanged. But a carrier itself carries no "modulation" (signal). Modulation is accomplished in this circuit by varying the frequency of the transmitted carrier in step with the desired signal.

In this case, the "desired signal" is the telephone conversation. The voices on the telephone line appear as varying voltages, easily shown on an oscilloscope. This changing-voltage signal is fed from the telephone line through capacitor C1 to the base of Q1, which is operating as a voltage-controlled oscillator.

The varying voltage at the base of Q1, bypassed and filtered by resistor R2 and capacitor C3, makes Q1 change its transmitted frequency. As the voltage at the base goes up and down, the transmitted frequency of Q1 shifts up and down—"frequency modulation"—which is exactly what an FM receiver is designed to detect and amplify.

Construction

Using the parts described in the parts list, you could build this project from scratch. However, if you were to wire this using a perforated board, you could easily run into unwanted oscillation from long leads, so the printed circuit board layout shown in Figure 4-2 is recommended. Using this layout, the parts locations are shown in Figure 4-3.

All of the parts are readily available except for L1, a small wire coil used in the tuning circuit. This can be made from 5$1/_2$ close-

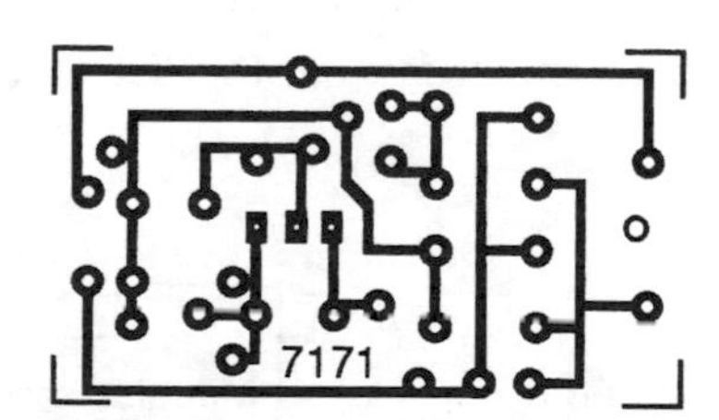

Figure 4-2: Printed circuit board layout diagram; board should be the exact size of this figure.

wound turns of #24 (0.020-inch diameter) enameled copper wire with a 5/32-inch inside diameter. (The parts kit provides this coil already wound.)

The modular cables connected to IN1, IN2, OUT1, and OUT2 are made from an RJ11 modular extension cable (modular plug on one end, modular jack on the other end) cut in half. The cut ends are separated from the sheath to reveal four wires: red, green, yellow and black. Only the red and green wires are used; the yellow and black wires are cut short.

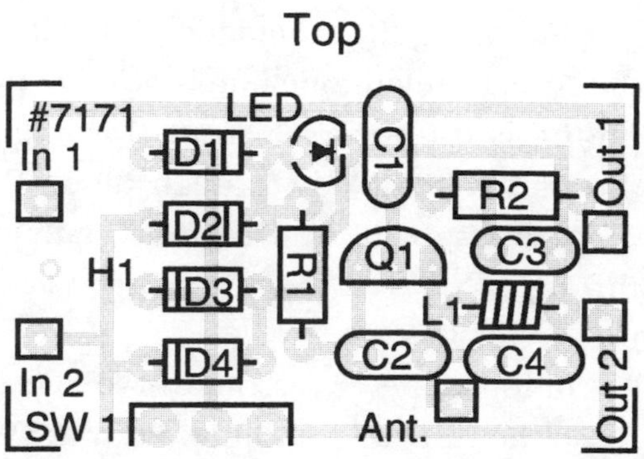

Figure 4-3: Parts location and placement on the printed circuit board.

Before going any further, you should decide where the telephone FM transmitter will be inserted into the phone line. You can connect either at the telephone wall jack or at the telephone itself.

If you decide to plug the telephone FM transmitter into the telephone wall jack, as shown in Figure 4-4, then the modular cable PLUG connects to the IN1 and IN2 terminals of the telephone FM transmitter, with the modular cable JACK connected to the OUT1 and OUT 2 terminals. The modular plug from the telephone then plugs into the telephone FM transmitter jack.

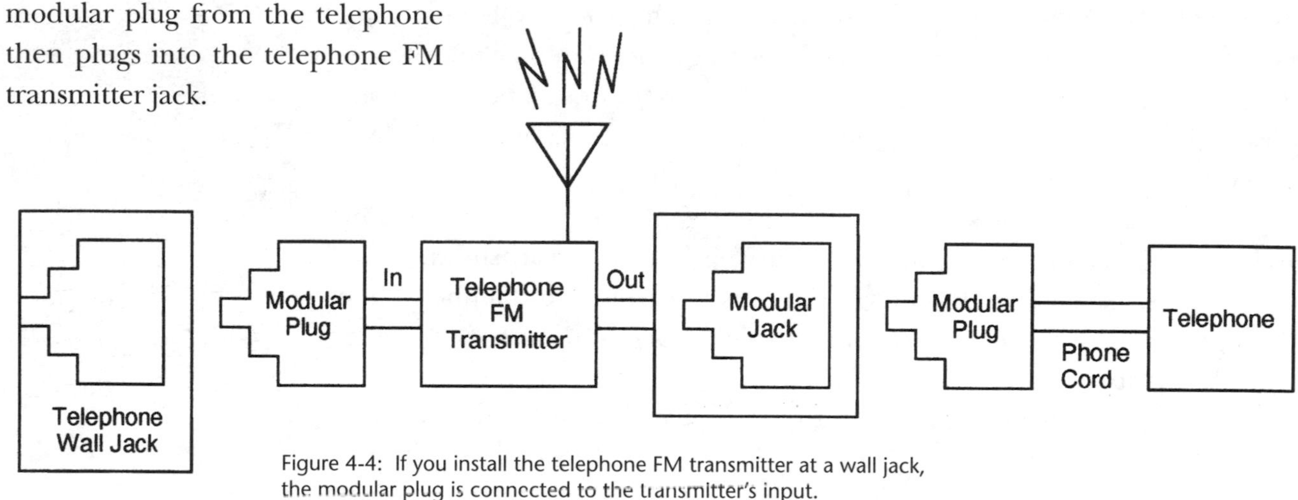

Figure 4-4: If you install the telephone FM transmitter at a wall jack, the modular plug is connected to the transmitter's input.

On the other hand, if you would rather have the telephone FM transmitter connected near the telephone itself, as shown in Figure 4-5, then the modular cable JACK is connected to the IN1 and IN2 terminals of the telephone FM transmitter, with the modular cable PLUG connected to the

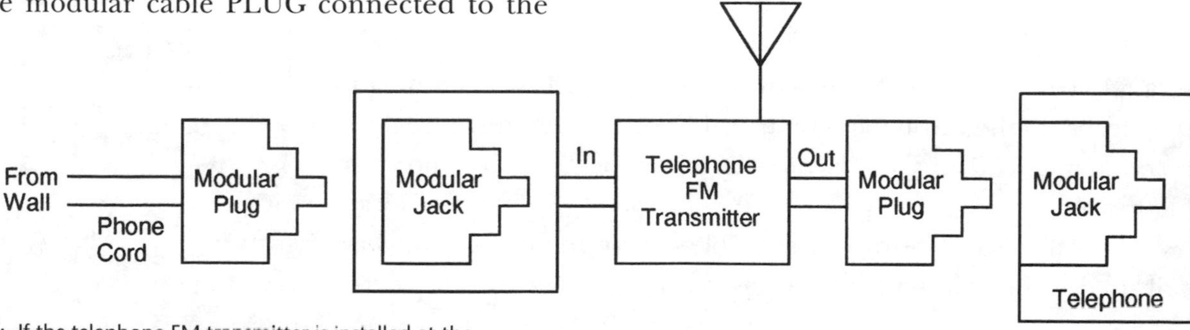

Figure 4-5: If the telephone FM transmitter is installed at the telephone, the modular jack is connected to the transmitter's input.

OUT1 and OUT2 terminals. The modular plug from the telephone wall jack phone cord then plugs into the telephone FM transmitter modular jack. In either case, Figure 4-6 shows that the modular cable green wires are connected to IN1 and OUT1, and the red wires to IN2 and OUT2.

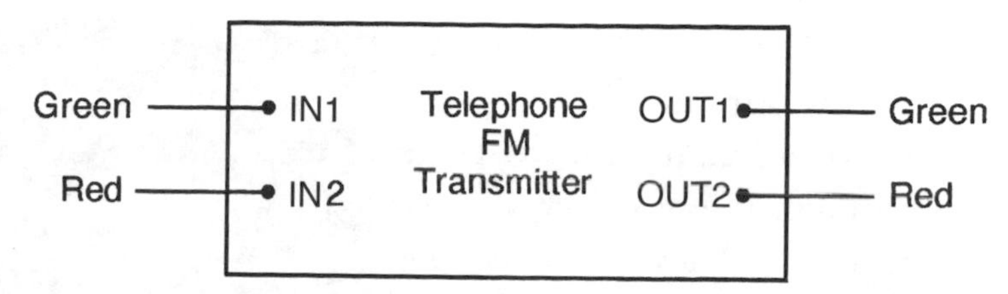

Figure 4-6: The green and red cable wires are connected to IN1, IN2, OUT1, and OUT2 as shown here.

Make sure you use the proper orientation in placing the diodes and the LED. The transistor must also be connected properly with respect to the base, emitter, and collector. The printed circuit board is designed for the lead arrangement of the 2N3906 transistor, shown in Figure 4-7.

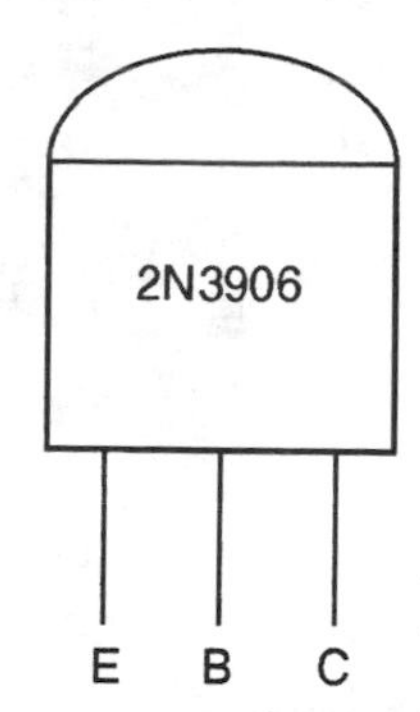

Figure 4-7: How to identify the leads of the 2N3906 transistor from its "flat" side.

The antenna is a short wire; about 6 inches long is adequate for a typical room-length transmission distance. A longer antenna will give you greater range, but will change the operating frequency, and may load the circuit so that it won't operate. Also, be aware that the greater the transmission range, the greater the chance someone next door—or down the block—will happen to tune their FM radio to your phone conversation!

Figure 4-8 shows the telephone FM transmitter assembled using the printed circuit board in the parts kit. As you can see, the completed circuit is small.

Figure 4-8: Here is the assembled printed circuit board. Note the parts placement, the cable connections, and the antenna wire running from the lower right.

Packaging

I prefer to have my projects packaged in such a way that the circuitry is protected and the project can be handled without unnecessary wires, switches, lights, and controls hanging loose. In this case, except for the external plug-in cables, the entire printed circuit board and all its parts are mounted inside a 35mm film container—the two-piece round plastic container used to hold a roll of 35mm film.

These containers are readily available from any film processor, since they usually accumulate from film developing and are thrown away! While most are opaque black, some are translucent, which would be preferred in this project since you'll then be able to see L1 light when the circuit is transmitting.

It is simple to drill, or use a hot soldering iron, to make holes through the plastic cover and bottom of the 35mm film container. Feed the wire end of one cable through the cover, and the other cable end through the container bottom, before wiring to the circuit board. Also, make a small hole to feed through the antenna wire. It turns out that the printed circuit board provided with the parts kit is an exact fit for the inside of a 35mm container! Figure 4-9 shows the results I achieved.

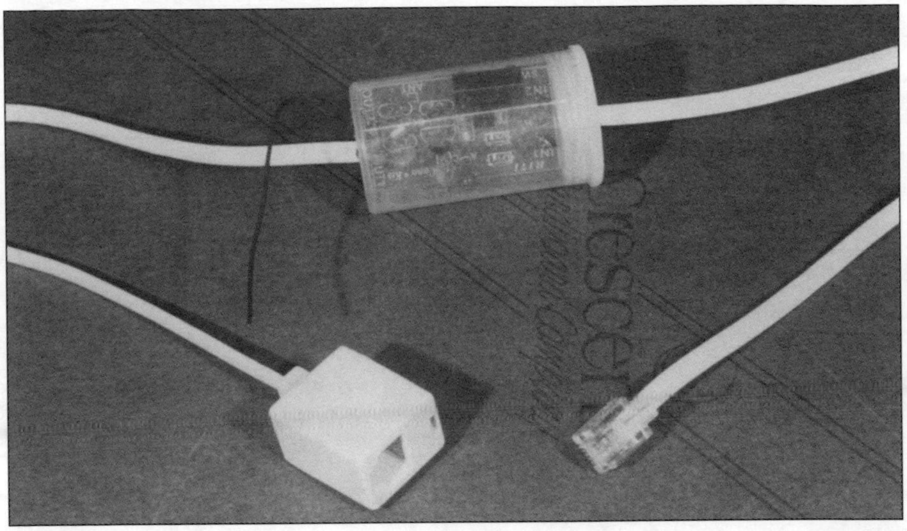

Figure 4-9: The assembled telephone FM transmitter can fit neatly into a standard 35mm film container, as shown here. The wire protruding at the left from the container is the antenna.

Installation

Insert the assembled telephone FM transmitter into the phone line as shown in either Figure 4-4 or Figure 4-5. Be sure the IN1 and IN2 terminals are nearest the telephone wall jack, NOT the telephone!

The installation shown in Figure 4-5 uses a standard 4-wire phone cord, with RJ11 modular plugs at both ends, between the telephone wall jack and the input to the telephone FM transmitter. It also assumes a telephone with a modular jack. If the telephone has a permanent phone cord terminating in a modular plug, you'll need a common telephone 4-wire in-line coupler to join it to the modular plug at the output of the telephone FM transmitter.

Once connected, lift the telephone handset and the telephone FM transmitter LED should light. If the LED does not, put the SW1 switch in its other position, and the LED should light. (If it still does not light, consult the following "Troubleshooting" section.)

When the LED is on, the unit should be transmitting a dial tone somewhere at the low end of the standard FM broadcast band. Tune a nearby FM receiver until you hear the dial tone, then dial a number, and you'll hear both sides of the conversation on the FM receiver.

In large cities, where the FM broadcast band is very crowded, you will probably encounter broadcast interference. The closer the FM receiver is to the telephone FM transmitter, the stronger the received signal, and the more likely it will overpower a broadcast station. However, if broadcast interference is a problem, you can increase the transmitting frequency of the telephone FM transmitter by SLIGHTLY spreading one or two turns of coil L1. Caution: spread the coils too far and you will be transmitting above the upper receiving frequency of your FM radio!

Incidentally, once the telephone FM transmitter is working, you can remove the LED from the circuit for greater transmitter output.

Troubleshooting

Chances of error are minimized if you use the printed circuit layout. The kit described at the end of the parts list includes a printed circuit board with the parts locations silk-screened on the component side, making errors even less likely. The most common problem in home-built electronic devices is bad soldering. Check all solder joints carefully; they should be shiny, not dull gray. The next most common problem is placing parts in the wrong location, especially resistors and capacitors. Check the values of each carefully, and make sure they are where they belong in the circuit.

In this circuit, the polarity (orientation) of the four diodes, the LED, and the 2N3906 transistor are critical. The diode and LED symbols show an arrowhead and a bar. The bar is the cathode, usually marked with a solid black or white band. The LED cathode usually has a flat spot at the base. (The transistor leads were shown in Figure 4-7.)

PARTS LIST

C1:	330 pF ceramic disc capacitor
C2:	12 pF ceramic disc capacitor
C3:	470 pF ceramic disc capacitor
C4:	22 pF ceramic disc capacitor
D1, D2, D3, D4:	1N914 or 1N4148 signal diode
LED:	Red light-emitting diode
L1:	Coil (see text)
Q1:	2N3906 transistor
R1:	180 Ω 1/8 W 5% carbon film resistor
R2:	12 kilohm 1/8 W 5% carbon film resistor
SW1:	SPDT slide switch
Cable:	4-wire telephone extension cable with RJ-11 modular plug at one end, modular jack at other end.
PC board:	Etched, drilled, and silk-screened (see Figure 4-2)

Other Uses

In addition to group listening to both sides of a conversation, you can record the conversation by holding a recorder near the speaker of the FM radio.

For a better recording, you can connect the earphone output of the receiver to the microphone input of a recorder, using a recording-level attenuator or attenuating dubbing cord, such as those available from Radio Shack and other electronics suppliers. However, be aware that recording a telephone conversation without the knowledge of the speaking parties is unlawful in most states.

NOTE: *At the time this book was being written, a complete kit of all parts was available as "Cana-Kit CK171 Telephone Conversation Transmitter" from Centerpointe Electronics, Inc., 5241 Lincoln Avenue, Cypress, CA 90630; toll-free order line was 800-272-2737. The price was $12.95 plus $5 shipping. Please remember that availability and pricing may have changed since publication of this book.*

CHAPTER 5

A Telephone Line Analyzer

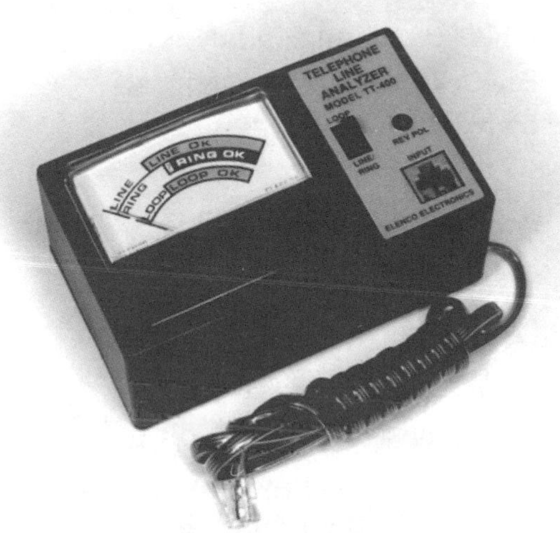

A telephone line analyzer allows you to check your telephone line for proper DC line voltage, AC ring voltage, and loop current and will also detect an open or short circuit in a detachable line cord. You can build it from scratch with parts from your "junk box" or from an inexpensive kit that contains all the parts, including a specially marked meter and a custom cabinet.

When your telephone quits working, it can be due to any number of different causes. If you are familiar with telephone operation and have a multimeter, you can track down most problems. But you can really simplify those tasks with the telephone line analyzer described in this chapter.

How Telephones Operate

Before you can properly use the telephone line analyzer, you need to know how the telephone system and individual telephones connected to it work.

The primary purpose of the telephone is to transmit and receive voice signals that allow two people with telephones to communicate. To be of practical value, the telephone must be connected to a switching network capable of connecting each telephone to many other telephones.

To accomplish this switching, each subscriber telephone is connected to the telephone company's "central office" by two wires referred to as the "local loop." A simplified diagram of this connection is shown in Figure 5-1.

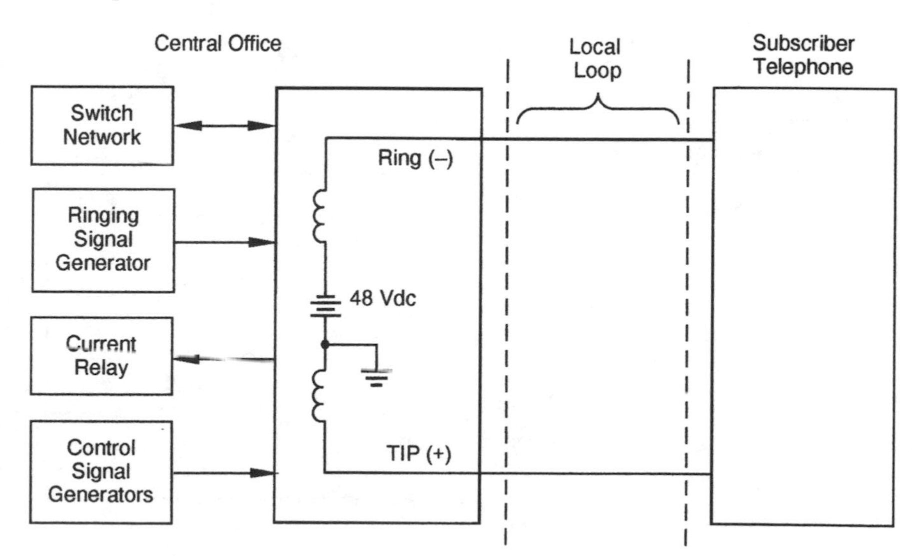

Figure 5-1: A simplified diagram of the central office/telephone connection.

The "tip" and "ring" designation of the + and – leads comes

from the days of the manually-operated switchboard. The tip of the plug that the operator used to connect telephones carried the + lead, and the ring immediately behind the tip carried the – lead.

Normally the phone is hung up; that is, "on hook." When you wish to place a call you merely pick up the telephone ("off hook") and a small current flows in the local loop. This current activates a relay in the central office indicating that service is being requested. When the central office is ready to accept the number being called, a dial tone is sent to your telephone. The dial tone stops when the first digit is dialed or pressed.

When you "dial" a number (rotary or pulse), tones or pulses are decoded at the central office. A path is established to the telephone being called. This path may be a simple wire connection to a telephone connected to the same central office, or it may go via wire, microwave link, or satellite to a telephone connected to a distant central office.

To signal the incoming call, a ringing signal is placed on the local loop by the central office of the called telephone. The ringing signal is a 90-volt alternating current, at 20 Hertz (Hz) per second, signal superimposed on the direct current voltage of the local loop. A ringing tone is also sent to the calling telephone. When the called party picks up the phone, the ring signal stops and voice communication is established.

Inside a Telephone

A simplified schematic diagram of a traditional mechanical rotary dial telephone is shown in Figure 5-2. The dialing mechanism here creates pulses. In modern telephones, tone dialing is used and many of the bulky parts of the old rotary telephone are replaced by transistors, integrated circuits, and piezoelectric buzzers.

When the hook switch (S1 in Figure 5-2) is open (on hook), no current flows in the local loop. The nominal 48 volts DC (VDC) from the battery in the central office appears on the tip and ring input to the telephone set. When the receiver is lifted (off hook), switch S1 closes and a current of about 20 to 120 milliamperes (mA) flows in the local loop. The resistance of the local loops drops the voltage at the telephone to about 6 VDC.

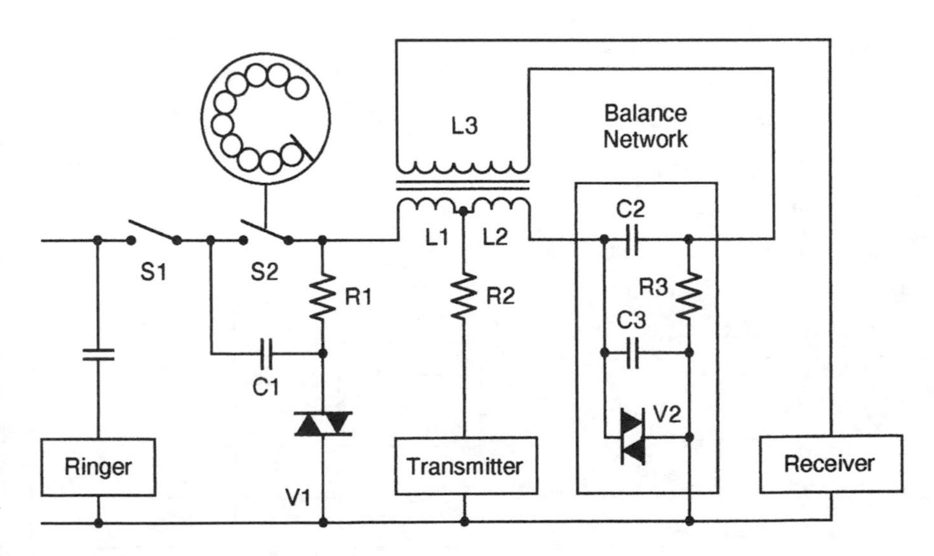

Figure 5-2: A simplified schematic of a typical rotary dial telephone.

Pulse and Tone Dialers

There are two types of dialers: pulse and tone. Pulse dialing is accomplished by the familiar rotary dial shown in Figure 5-2. The dial is rotated to the stop and then released. A spring in the dialer returns the dial to its null position. As the dial returns, the dial switch, S2, opens and closes at a fixed rate. This switch is in series with the S1 hook switch (which is now off hook, and therefore closed.)

Each opening of the dial switch interrupts the current in the local loop, thus generating a series of pulses, as shown in Figure 5-3. The number of current interruptions corresponds to the digit dialed. For example, dialing a 0 sends ten pulses.

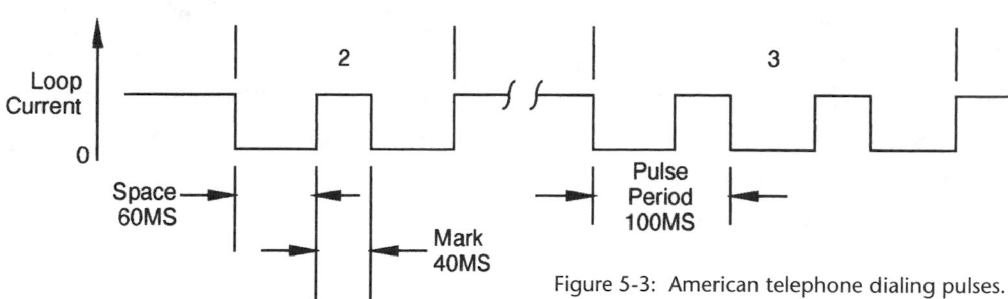

Figure 5-3: American telephone dialing pulses.

The dial pulses are sent at a rate of 10 pulses per second, so there are 100 milliseconds (ms) between pulses. Each pulse consists of a "mark" interval during which loop current flows, and a "space" interval indicating loop current interruption. In the United States, the mark interval is 40 ms and the space interval is 60 ms, giving a mark/space ratio of 40/60. In Europe, the mark/space ratio is usually 33/67.

Tone dialing is accomplished using a keyboard of 12 keys arranged in four rows and three columns, as shown in Figure 5-4. Seven distinct frequencies are generated within the telephone. The lower frequencies of 697, 770, 852 and 941 Hz are associated with rows R1 through R4. The higher frequencies of 1209, 1336 and 1477 Hz are associated with columns C1 through C3.

To send any digit, two frequencies are sent to the central office simultaneously. For this reason, this method of dialing is referred to as "dual-tone multi-frequency," or DTMF. This is accomplished by having any of the 12 keyboard buttons mechanically close two switches simultaneously. For example, if the "2" key is pressed, then switches SC2 and SR1 are both closed simultaneously, generating tones at 697 Hz and 1336 Hz. The "column" switches SC1, SC2, and SC3 are ganged to the "row" switches SR1, SR2, SR3 and SR4, as shown in Figure 5-4.

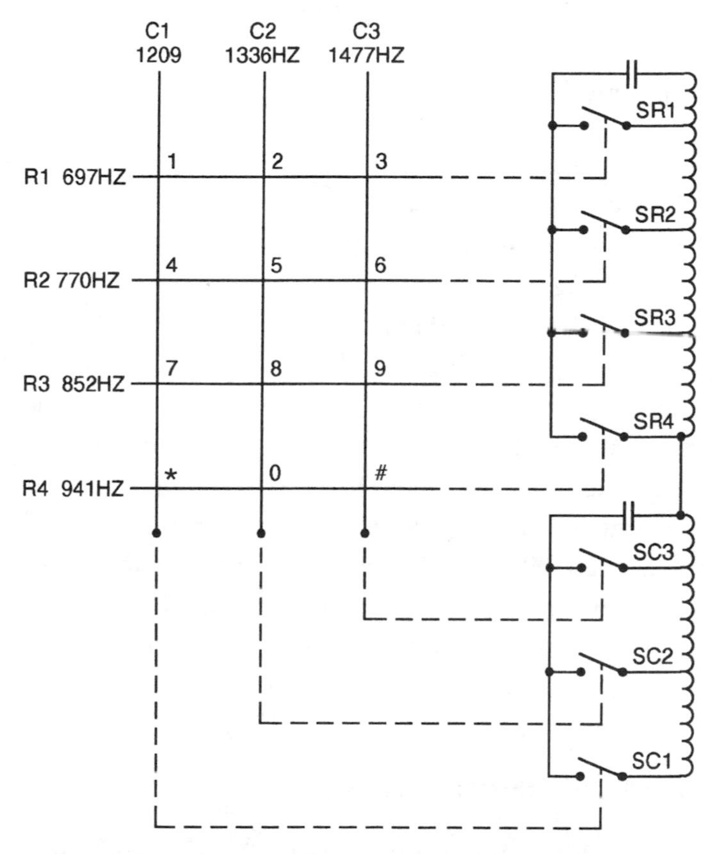

Figure 5-4: Dual-tone multifrequency (DTMF) dialing system.

The Telephone Transmitter

Older telephones used a carbon granule transmitter consisting of a metal diaphragm and a metal case insulated from each other, as shown in Figure 5-5. The case is filled with carbon granules. Wires are connected to the case and diaphragm, and a current is put through the carbon granules.

When you speak into the transmitter, the sound waves of your voice strike the diaphragm and cause it to vibrate. This causes the carbon granules to compress and expand. When compressed, the resistance of the carbon granules is less than when expanded. The change of resistance causes a corresponding change in the current. The current thus varies in step with the sound waves of your voice.

In newer telephones, the carbon granule transmitter is replaced by an electret or other type of microphone. An electret microphone is made up of a capacitor with a dielectric material that holds a permanent electric charge. Sound waves striking a plate of this capacitor cause the plate to vibrate and thus generate a small voltage across the capacitor. This voltage is amplified by a field effect transistor (FET) mounted inside the microphone. The signal from the microphone is then amplified before being sent to the central office via the local loop.

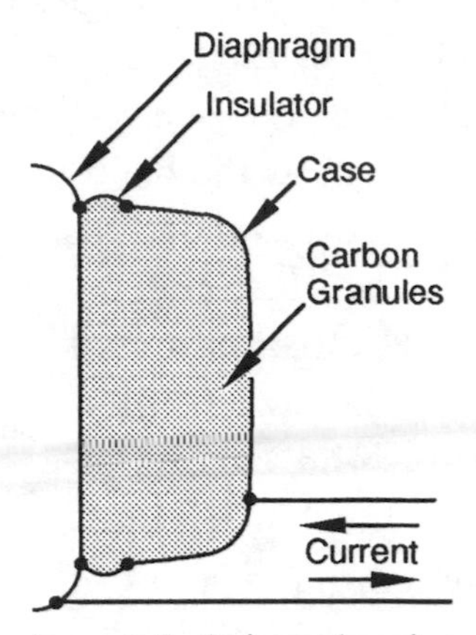

Figure 5-5: Carbon microphone in a typical telephone.

The Telephone Receiver

There are several different types of receivers. In principle they work the same as the speakers in your radio and TV. The speaker consists of a small coil attached to a diaphragm. The coil is mounted over a permanent magnet as shown in Figure 5-6.

Coil current in one direction causes the coil and diaphragm to be repelled from the permanent magnet. Coil current in the other direction causes the coil and diaphragm to be attracted to the permanent magnet. If a current of audio frequency is sent through the coil, the diaphragm vibrates and generates sound waves in step with the current. Thus, if the current from the transmitter is sent through the coil, the sound produced will duplicate the sound striking the transmitter.

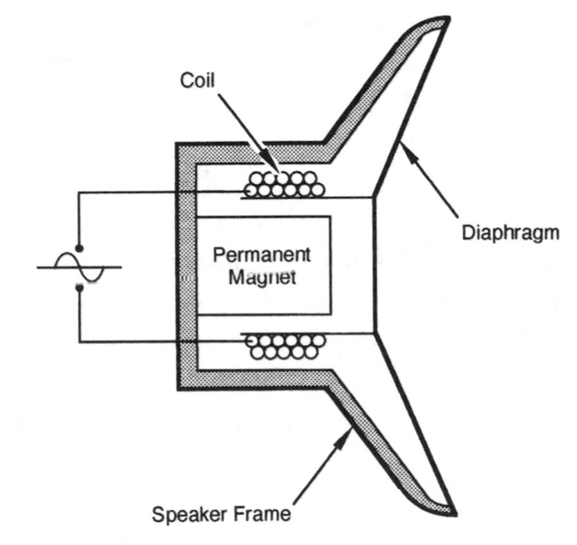

Figure 5-6: Typical permanent magnet telephone receiver.

The Telephone Ringer

As previously shown in Figure 5-2, the ringer is connected across the tip and ring inputs in series with a capacitor to block the on-hook 48 VDC line voltage. A typical mechanical bell ringer consists of a permanent magnet attached to an armature, as shown in Figure 5-7. When an alternating current of 20 Hz is passed through the coils, the armature is alternately attracted to one coil and then the other. The hammer attached to the armature thus strikes one bell and then the other to produce the ringing sound.

In electronic telephones, the hammer and bell ringer are replaced by a piezoelectric buzzer driven by an electronic oscillator. A piezoelectric material changes its dimensions when a voltage is applied. When the oscillator is running, it applies a 3,000 to 4,000 Hz AC voltage to the piezoelectric buzzer. The buzzer changes its dimensions and produces a 3,000 to 4,000 Hz sound. Since the oscillator is turned on only during one half of each cycle of the 20 Hz ring signal, the buzzer produces sound that is switched on and off 20 times each second.

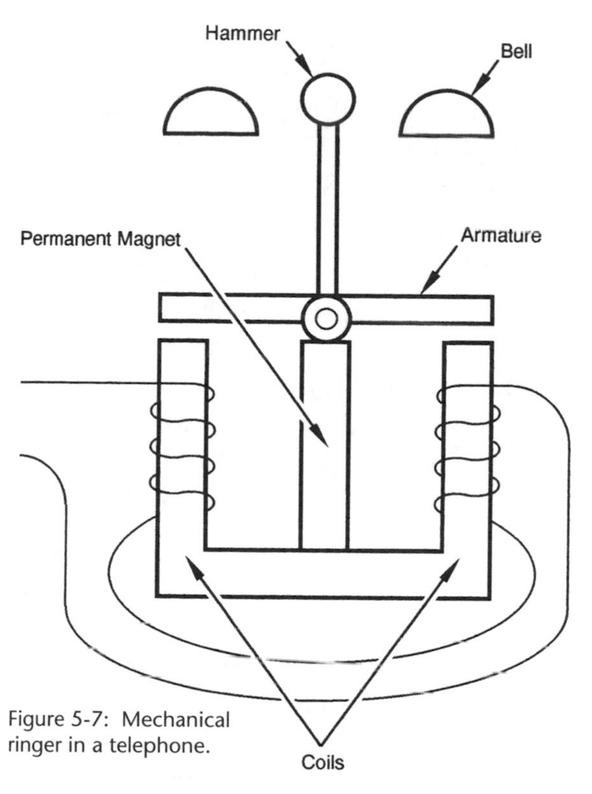

Figure 5-7: Mechanical ringer in a telephone.

Induction Coil/Balance Network

When transmitting and receiving is done over the same two wires, the problem arises that current from the transmitter flows through the receiver. The speaker then hears his own voice—called the "sidetone"—from the receiver. Too much sidetone may be objectionable to the speaker and cause him to speak too softly. However, a small amount of sidetone is desirable to keep the telephone from sounding dead.

The balance network of older telephones, as previously shown in Figure 5-2, requires some explanation. The induction coil and balance network limit the sidetone. The impedance of the balance network approximately matches the impedance of the local loop. Thus about half of the current from the transmitter flows through L1 and the local loop, and the other half flows through L2 and the balance network. The currents in L1 and L2 induce voltages in L3 of opposite polarity, thus limiting voltage across the receiver to an acceptable level.

When receiving a signal from the local loop, the currents in L1 and L2 induce voltages in L3 of the same polarity and these voltages combine to drive the receiver. Newer telephones perform this function electronically.

Telephone Line Analyzer Circuit Description

Now that you have some understanding of how a telephone system works, let's take a look at the circuit operation of a telephone line analyzer. Figure 5-8 shows the schematic for such a device (this is the same circuit used in the Elenco model TT-400K telephone line analyzer kit discussed later in this chapter). Basically, Figure 5-8 shows a circuit consisting of a specially-marked one milliampere meter movement with the necessary parts to read telephone line current and voltage.

As you may already know, a milliampere meter can read current greater than the basic meter movement by using a parallel path for current around the meter. A meter can also read voltage by adding series resistance to reduce the current flow to the meter's basic movement. By the proper selection of resistor values, you control the full scale meter readings.

The plug at the end of the telephone cord, P1, plugs into a wall telephone jack. If wired properly (green is DC positive, red is return), this feeds positive voltage to J1, with the return through J2.

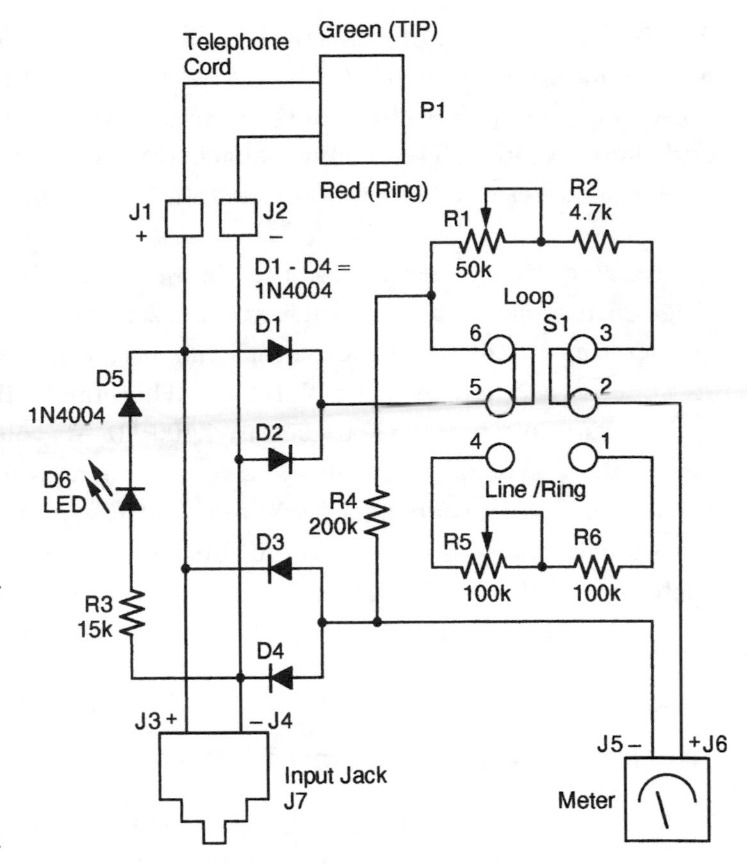

Figure 5-8: Schematic for the telephone line analyzer.

Assuming the wall jack is properly wired, and switch S1 is in the LOOP position, as shown, and nothing is plugged into input jack, J7, the local loop current from the central office flows through diode D1, through terminals 5 and 6 of S1, through potentiometer R1 and resistor R2, through terminals 3 and 2 of S1, through the DC meter, and returns to the central office through diode D4.

But notice resistor R4. This allows some current to go back to J2 through D4 without passing through the meter. This is known as a "shunt." The full-scale meter reading is 60 mA, but by the proper selection of shunt resistor value, and setting potentiometer R1 to the proper value on calibration (as described later), the beginning of the LOOP OKAY marking on the meter face is about 20 mA. Diodes D2 and D3 allow the circuit to function even if the wall jack polarity is reversed. (We'll discuss this later in this chapter.) Also, D3 and D4 prevent positive voltage from appearing on the meter negative terminal with either wall jack polarity.

If switch S1 is placed in the LINE/RING position, current flows through D1, through switch terminals 5 and 4, through potentiometer R5 and resistor R6, through switch terminals 1 and 2, and then through the meter, returning through D4. By properly setting R5 during calibration, the full-scale meter reading is 120 VDC, making the beginning of the LINE OK marking on the meter face about 40 VDC.

When the phone line is receiving a ringing signal, about 90 volts of alternating current is applied on top of the normal 48 VDC voltage 20 times a second. The negative half-cycles of this alternating current cause the LED to flicker. The alternating current is rectified by diodes D1 and D2 to provide pulsating DC, which is conditioned by the meter circuitry to fall into the ? or RING OK meter scale readings. The stronger the ring signal strength, the higher it will read on the meter scale.

Diode D5 protects the cathode of LED D6 from the approximate positive 48 VDC at J1, which would destroy the LED. If the wall jack is wired in reverse (red DC positive, green return), then current flows through R3, D6, and D5, and the LED turns on. At the same time, diodes D2 and D3 allow the meter current to flow.

Construction

If you build the telephone line analyzer using the TT-400K kit described at the end of this chapter, assembly is detailed in a well-illustrated assembly manual. The meter supplied with the kit has a custom marked face, and a faceplate identifies the switch positions. The kit also includes an etched, drilled and silk-screened printed circuit board, and a custom cabinet, both greatly increasing ease of assembly. The kit manufacturer, Elenco, will not supply individual parts—only the complete kit. If you want to build the unit from scratch, be aware that just the milliammeter—without the special markings—may cost more than the entire kit!

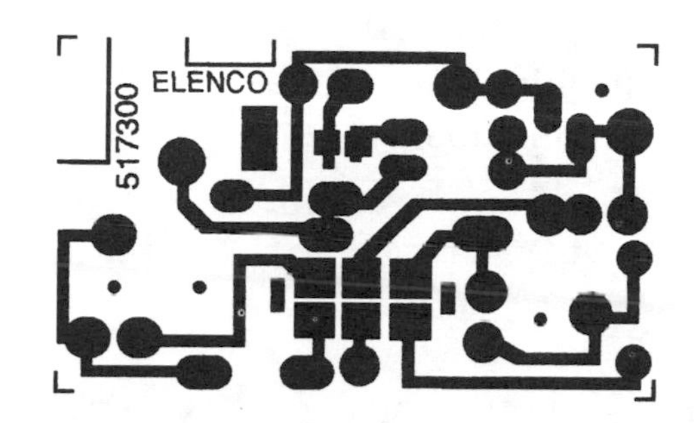

Figure 5-9: Printed circuit board layout.

Figure 5-9 shows the printed circuit layout, Figure 5-10 shows the parts layout and the special holes in the printed circuit board, Figure 5-11 is the marking of the meter face, and Figure 5-12 is the faceplate. If you don't build from the kit, you'll have to furnish your own cabinet and drill all holes to mate with the parts.

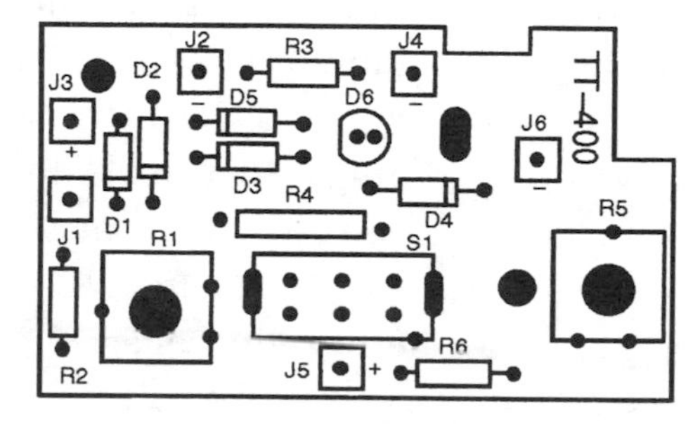

Figure 5-10: Parts layout and placement on the printed circuit board.

Figure 5-11: Meter face template.

Instead of making a printed circuit board, you could build the circuitry on a perforated board, wiring components directly following the schematic. This could be a daunting task, since you'd need to provide access to the potentiometers for calibration, and you'd have to find a way to securely mount the meter and the input jack, J7. Figure 5-13 shows the interior of the unit

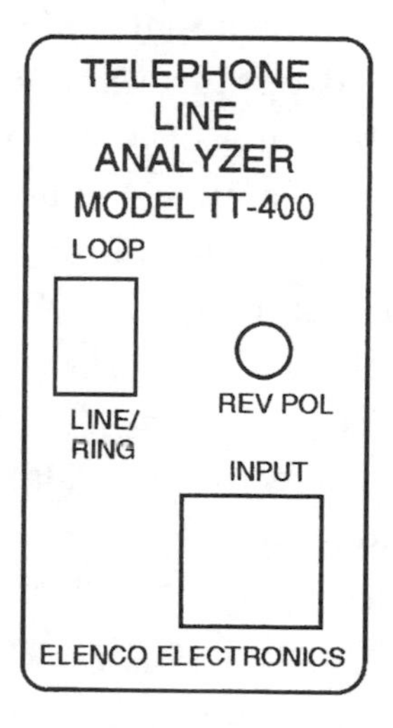

Figure 5-12: Faceplate for the telephone line analyzer.

when a printed circuit board is used for assembly. As you can see, construction and mounting of the circuitry is greatly simplified with this approach.

The parts list shows the sources for those parts that are hard to find. Most parts are available from Mouser and other electronic suppliers. However, keep in mind the availability of these parts (and their stock or catalog numbers) may have changed since this book was printed. If you use the parts kit described at the end of this chapter, you get a more professional looking unit, as shown in Figure 5-14.

Figure 5-13: Interior view of the unit when the printed circuit board is used.

Calibration and Troubleshooting

Once the unit is assembled, you need to calibrate it for proper meter readings. To calibrate the LINE/RING voltage scale, set S1 to the LINE/RING position, and apply 20 VDC. Adjust R5 until the meter needle is at one-sixth full scale (the last line of the letter N as shown in Figure 5-15). Reverse the voltage at J1 and J2, making J2 positive. The LED should light brightly and the meter reading should drop.

To calibrate the LOOP current scale, set switch S1 to LOOP. Using a source of variable voltage, or an appropriate dropping resistor with a fixed voltage source, apply 20 mA of direct current between J1 and J2, with J1 positive. With a milliammeter in series reading 20 mA, adjust R1 until the meter needle is just at the beginning of the LOOP OK scale.

If something isn't working, it is usually due to a bad connection or solder joint, a part in "backwards," or (rarely) a defective part. Make sure all diodes (and the LED) are properly oriented, that the incoming cable and jack J7 are wired properly, that the meter connections are not reversed, and that all connections are good.

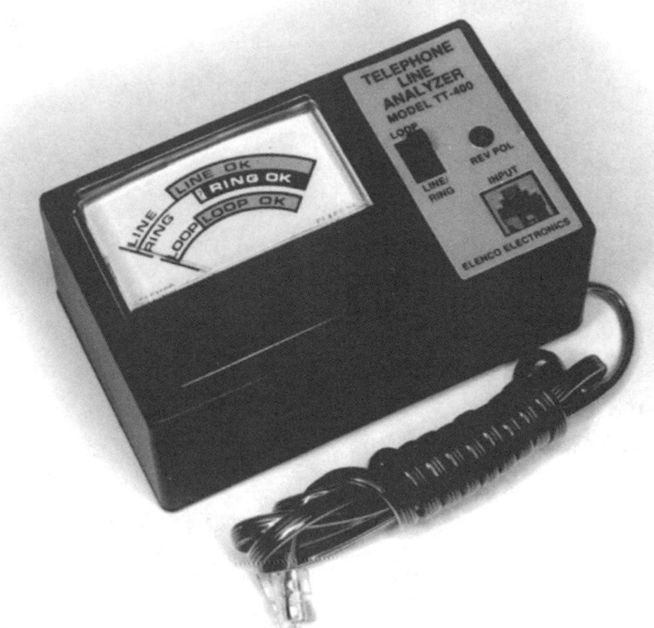

Figure 5-14: The completed telephone line analyzer with the parts I used.

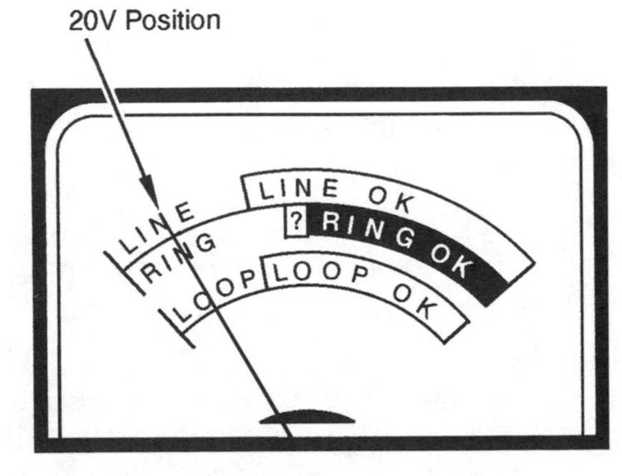

Figure 5-15: The calibration point for the line voltage.

Using the Telephone Line Analyzer

Four significant tests can be performed with the telephone line analyzer: line/polarity test, line cord test, ring test, and loop test.

The easiest is the line/polarity test. Set the S1 switch to LINE/RING and plug the phone cord from the telephone line analyzer directly into a telephone wall jack. If all phones on the line are hung up, this measures the local loop on-hook voltage, moving the meter needle into the LINE OK area of the face.

The polarity part of this test is just a matter of looking at the LED. The LED should NOT light! If it does, the wall jack is wired with reverse polarity for the red and green wires; green should be positive.

Now plug a telephone into the telephone line analyzer's INPUT jack. With all phones on the line hung up (on hook), nothing should change. Lift the receiver and the meter needle should drop to near zero (actually, about 6 VDC). If the meter reads below the LINE OK area with all phones on the line hung up, the telephone plugged into the telephone line analyzer, or an extension phone, may be defective. Unplug them one at a time to see if this brings the meter reading into the LINE OK area. If not, the problem is in the telephone line itself.

For the line cord test, set S1 to the LINE/RING position. Plug one end of the line cord to be tested into the telephone wall jack and the other end into the telephone line analyzer's INPUT jack. If the wall jack is wired properly, the LED should light (since the input jack, though marked INPUT, is actually an OUTPUT to a telephone, with red and green wires reversed). The meter should read in the LINE OK area. Jiggle the line cord. If the meter needle is erratic or drops near zero, the line cord is bad.

To perform the ring test, set S1 to the LINE/RING position, plug the telephone line analyzer into the telephone wall jack, and plug a telephone into the telephone line analyzer INPUT jack. All telephones on the line should be hung up. The meter should read in the LINE OK area. Now either call this phone number from another phone line, or have a friend call this phone number. When the phone rings, the LED should flicker, and the meter needle should increase about one-eighth inch up scale to the RING OK, or at least the "?" area of the meter scale.

If the meter does not increase, or the phone does not ring, you may have too many extension phones or devices on the same phone line. Each device has an REN (ringer equivalence number) that should be indicated on its nameplate. The total of all RENs (just add them) should be 5.0 or less.

To perform the loop test, put switch S1 into the LOOP position to measure local loop current. (Note: Anyone calling you while the telephone line analyzer is connected to the wall jack with the switch in the LOOP position will get a busy signal!) Plug the telephone line analyzer line cord into the telephone wall jack. The meter should read in the LOOP OK area. Plug a telephone into the telephone line analyzer INPUT jack. With all phones on-hook, nothing should change. If the meter needle drops, remove phones from the line until it's back to normal. Be aware that sometimes high demand from the central office can lower the local loop current. If it seems low, check it several times at 20 minute intervals.

NOTE: *At the time this book was written, a complete telephone line analyzer kit including all parts, a specially-marked meter, printed circuit board, cabinet, wire, solder and detailed assembly manual was available from: C & S Sales, Inc., 150 West Carpenter Avenue, Wheeling, Illinois, 60090; toll-free order number (800) 292-7711. The stock number of the kit was TT-400K and the price was $19.95 plus $5 shipping. Completely assembled units were also available. Please remember that availability and pricing may have changed since publication of this book.*

PARTS LIST

D1-D5:	1N4004 1 amp 400 PIV diode
D6:	Red light emitting diode
P1:	Telephone cable assembly (Western Electric D4QK-7, Mouser 154-3003, or equivalent)
J7:	Telephone modular Jack (Western Electric 623K4, Mouser 154-UL623K4, or equivalent)
R1:	50 kilohm 1/4 W potentiometer (Mouser 531-PT15B-50K)
R2:	4.7 kilohm 1/4 W 5% resistor
R3:	15 kilohm 1/4 W 5% resistor
R4:	200 Ω 1 W 5% resistor
R5;	100 kilohm 1/4 W potentiometer (Mouser 531-PT15B-100K)
R6:	100 kilohm 1/4 W 5% resistor
S1:	DPDT switch
M1:	1 milliampere DC panel meter (Mouser 541-MSQ-DMA-001)
Miscellaneous:	
	Wire, solder, cabinet, circuit board

CHAPTER 6

Two Simple Electronic Sirens

Here are two simple siren designs. One is a low-power wailing siren, and the other is a more powerful siren that either warbles or produces a single tone. These circuits are an experimenter's delight, since changing the value of a single resistor or capacitor can have a BIG effect on the circuits' operation.

The Wailing Siren

Although this "wailing siren" uses very few parts, it duplicates the upward and downward wail of a police siren. It is an experimenter's delight, since you can substitute some parts for different results. For example, you can construct it with two output volume levels—roughly 150 milliwatts or 2 watts—and couple it to a PA (public address) system for even greater power.

You can build the wailing siren from parts you probably already have in your junk box, or you can buy an inexpensive kit of most of the parts, including an etched and drilled printed circuit board. Ordering information will be given later in this chapter.

Circuit Description

The schematic of the wailing siren is shown in Figure 6-1. It will operate with a voltage from 3 to 12 volts DC. The low-power version is easily powered by a standard 9 volt transistor radio battery, since it draws a peak of less than 10 milliamperes from the battery at that voltage.

To begin with, ignore resistor R3 as if it were not there; I'll discuss R3 later. With the pushbutton switch open, as shown, transistors Q1 and Q2 are not conducting at all, and capacitors C1 and C2 have no charge. When switch S1 is pressed, electrolytic capacitor C2 starts to charge through resistor R2, but very little current is available through resistor R1 to the base of transistor Q2.

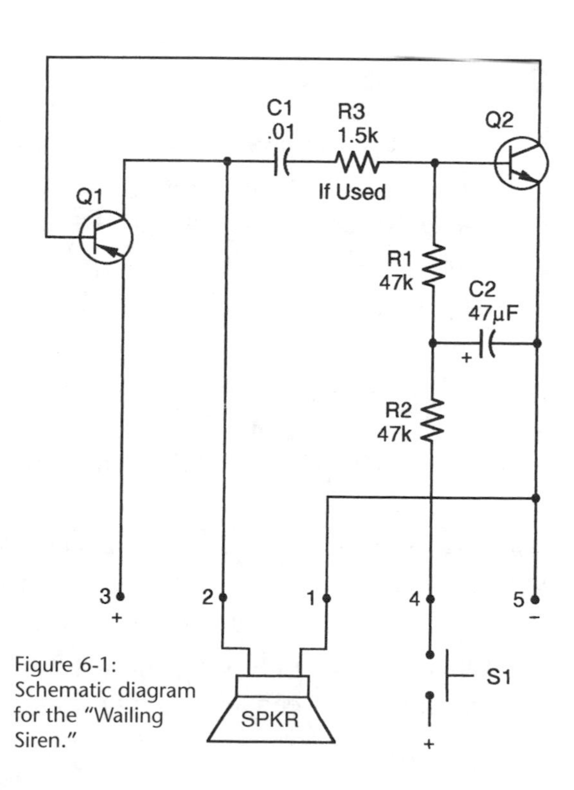

Figure 6-1: Schematic diagram for the "Wailing Siren."

As C2 charges, the voltage at R1 increases and allows current to flow through the base of the NPN switching transistor Q2, turning it on.

As soon as Q2 starts to conduct, it biases the base of PNP power transistor Q1, allowing it to conduct. This allows the voltage at terminal 2, one of the speaker terminals, to rapidly rise to almost the supply voltage. This rapid rise passes a positive pulse through disk capacitor C1 to the base of Q2, throwing it into full conduction. The positive pulse also is sent to the speaker.

When Q2 is thrown into full conduction, the amount of base current needed by Q2 is greater than the total current that can be supplied through the almost-100K series resistance of R1 and R2, and from the positive charge on C2. Although the voltage on C2 is constantly increasing, resistor R1 limits the capacitor current that is supplied to the base of Q2.

After the pulse through C1, the voltage at the base of Q2 quickly drops below the 0.7-volt bias needed, turning off Q2. This in turn shuts off Q1. C1 discharges through the speaker. Capacitor C2 has continued to charge—so long as S1 is held closed—so it is at a slightly higher voltage, allowing the conduction of Q2 and Q1 a bit sooner. Each cycle takes a little less time—as C2 continues charging, turning Q2 on sooner each cycle—so the frequency rises until it reaches a stable frequency.

The rate of charge and discharge determines the frequency of the tone generated through the speaker. Since the voltage on C2 is constantly rising as long as switch S1 is held closed, the frequency increases until it reaches the point where the charge on C2 no longer increases. When S1 is opened, C2 discharges a little with each cycle, taking it longer and longer to trigger Q2 each time; thus, the frequency drops.

Construction

The parts list shows all the parts needed for this project, none of which are particularly critical, although you certainly want a power transistor capable of several watts for Q1. Transistor Q2 can be a small switching NPN type.

This is an oscillator circuit, and therefore subject to erratic operation depending on lead lengths and placement between parts of the oscillator circuit. I'd recommend you build this project on the printed circuit board, such as shown in Figure 6-2 (and supplied with the parts kit). Parts placement is shown in Figure 6-3. Keep all component leads short.

Make sure the polarity of C2 is as shown, and that the emitter, base, and collector of each transistor are correctly placed. For Q1, typically in a TO-220 package with a metal tab, the tab should be closest to resistor R2. For Q2, typically in a TO-92 package with a flat side, the flat side should face Q1. Of course, if you're using some junk box transistor, it's up to you to determine the E-B-C lead arrangement, since they vary.

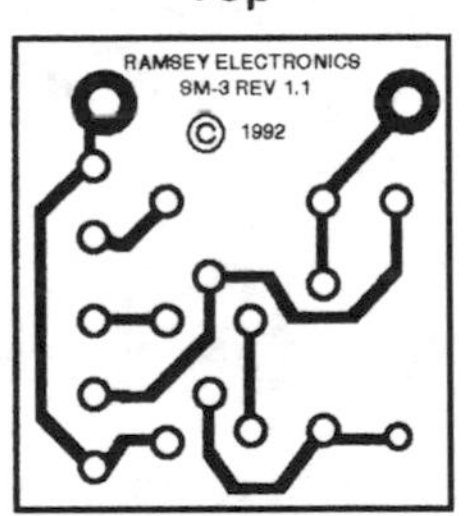

Figure 6-2: Circuit board layout for the "Wailing Siren" (actual size).

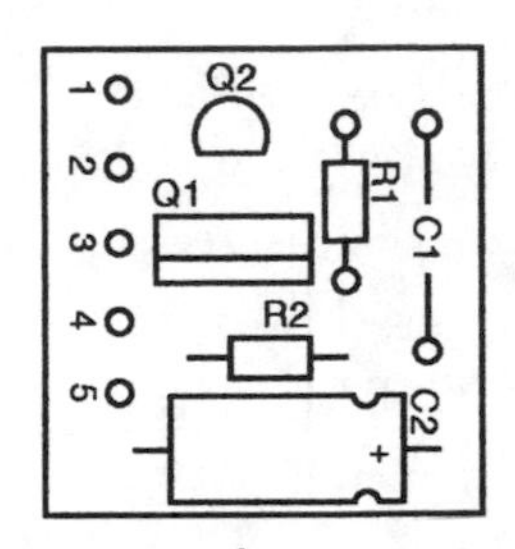

Figure 6-3: Parts placement for the "Wailing Siren."

Testing

If you just solder bare wire leads temporarily to the points 1–5 on the schematic, you can test out the completed circuit with clip leads. External connections to these points have little or no effect on the oscillator operation.

Connect positive voltage—a common 9-volt battery is fine—to Point 3, with the negative return to Point 5. The speaker can be any type with an impedance of from 3 to 45 Ω, but a horn type will give the most directed volume. Connect it between Points 1 and 2.

Now all you do is momentarily connect Point 3 to Point 4, simulating a pushbutton switch. You should hear a rising tone through the speaker. It won't be ear-shattering. Disconnect Point 3 from Point 4 and the tone should be descending in frequency. When you "close the switch," the tone goes up; open it and the tone goes down. Simple enough!

Troubleshooting

But what if it doesn't work? First check component orientation for the transistors and C2, then be sure all solder joints look okay. If everything seems to be right, you could still have a poorly soldered connection. Try gently bending the power transistor, Q1, for example, to see if the leads are firmly soldered. If the leads move at all, you probably don't have a good connection. Sometimes the leads are dirty and don't solder well. This happened to me several times with Q1. Also, when testing, I had the leads at Points 1 to 5 twist loose from the PC board, despite many years of soldering experience!

Modifications

While this circuit fairly begs for modifications, they have to be done neatly. When I tried making some component changes, I used long clip leads with resistor and capacitance substitution boxes, and the results were highly erratic! I had problems with excessive current draw (probably high-frequency parasitic oscillations above hearing range), and horribly raucous bursts of sound. I suggest each part to be substituted or added should be soldered in place on the PC board.

Resistor R3 can be added in series with C1 to produce a louder sound. A 1.5 K resistor is included in the parts kit for this purpose; you might wish to try other values. Adding R3 definitely increases the volume, but it also increases the current drain well beyond the amount a common 9-volt battery can supply for any reasonable time.

If you want the extra volume, add R3, but use either a 6-volt lantern battery or an external power supply capable of handling about 200 milliamperes. To be sure, hook an ammeter in series with the power supply you intend to use, and measure the current to see if you have what you need.

If you want even MORE volume, couple the output of the low-power wailing siren to a PA system or other amplifier. Instead of connecting a speaker between Points 1 and 2, use a 15 Ω 2 W resistor and 0.1 μF capacitor as shown in Figure 6-4. Plug this audio output to the microphone or auxiliary input jack of the amplifier. Start with a low volume setting on the amplifier, since the output voltage of the wailing siren is considerably more than a typical microphone. You may even need an attenuating plug or cable, as sold by Radio Shack and other electronics suppliers, to keep from overpowering the amplifier if it only has a microphone input.

The other part you may want to experiment with is electrolytic capacitor C2. Change this from a value of 47 μF to 5 μF or 10 μF, and you'll find the wailing occurring at a faster rate.

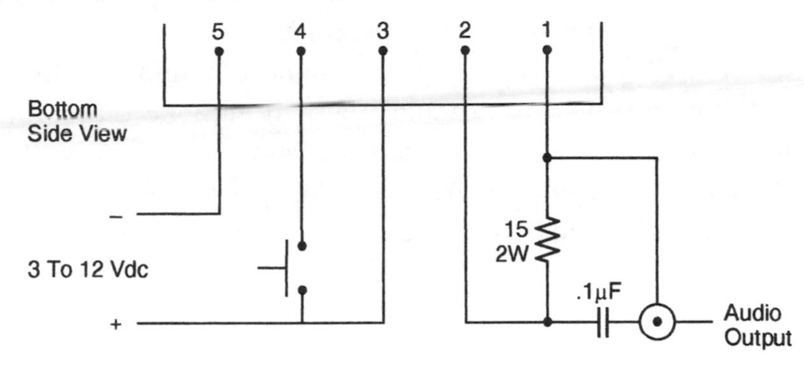

Figure 6-4: Coupling siren output to an external amplifier or PA system.

Packaging

For the low-powered version of the wailing siren, any small plastic enclosure will do if you use a typical small speaker of about 2.25 inches in diameter—the kind commonly found in pocket radios and recorders. The higher powered version needs a larger speaker and battery to handle the roughly 2 W output, so you have the choice of either a larger enclosure, or a smaller enclosure with a plug-in speaker.

My junk box happened to contain the empty case from a Citizen's Band walkie-talkie I bought some time in the early 1950s. The "Spacephone" was so poor in performance I junked the circuit but saved the case, thinking I might use it sometime. Figure 6-5 shows how perfectly the circuit board, speaker, battery, and pushbutton switch fit in this 40-year-old case! Figure 6-6 shows the outside view when this housing is used.

Figure 6-5: The "Wailing Siren" packaged inside an old toy walkie-talkie case. The pencil points to the power transistor.

Figure 6-6: Exterior view of the packaged "Wailing Siren."

WAILING SIREN PARTS LIST	
* C1:	0.01 µF ceramic disc capacitor
* C2:	47 µF electrolytic capacitor, 15 WVDC
* R1, R2:	47 kilohm carbon resistor, 1/4 W
* R3:	1.5 kilohm carbon resistor, 1/4 W (optional)
* Q1:	PNP power transistor, TO-220 case, 5-watts or better
* Q2:	2N3904 NPN switching transistor
SPKR:	Speaker (see text)
S1:	Pushbutton switch, normally open
Battery :	See text
* PC board:	Optional; see text

At the time this book was being written, a kit which included all the parts marked with an asterisk (*) above plus an assembly manual was available from Ramsey Electronics, Inc., 793 Canning Parkway, Victor, NY, 14564; phone 1-800-446-2295. Their stock number was SM-3. The cost per kit was approximately $4, although there was a shipping charge of $4.95 per order and an additional $3 charge on all orders under $20. Availability, pricing, and ordering policies may have changed since publication of this book.

The Mad Blaster Siren

This "mad blaster" siren/CPO is a relatively simple project that uses only nine parts with an external power supply and speaker to produce a very loud warbling siren or single tone that can be easily heard in an auditorium or large outdoor area. With an external telegraph-type key, it can also be used as a code practice oscillator (CPO). All the parts are common and available from various electronic part vendors. A parts kit is also available and will be described later in this chapter.

The mad blaster can be packaged in a pocket-sized unit. The circuitry consists of only one 4011 integrated circuit (IC), two transistors, and a few resistors and capacitors. It is easily powered by a car battery or other source of 9 to 12 volts DC at up to one ampere. As you'll see later in this chapter, I built the unit, with several options, into a small plastic case that was originally designed to hold 36 color slides!

Circuit Description

The "mad blaster" circuitry, with various options, is shown in Figure 6-7. The CMOS 4011 integrated circuit (U1) consists of four 2-input NAND gates used to form two square wave oscillators operating at different frequencies. The combined oscillator output is amplified through two transistors and fed to an external speaker.

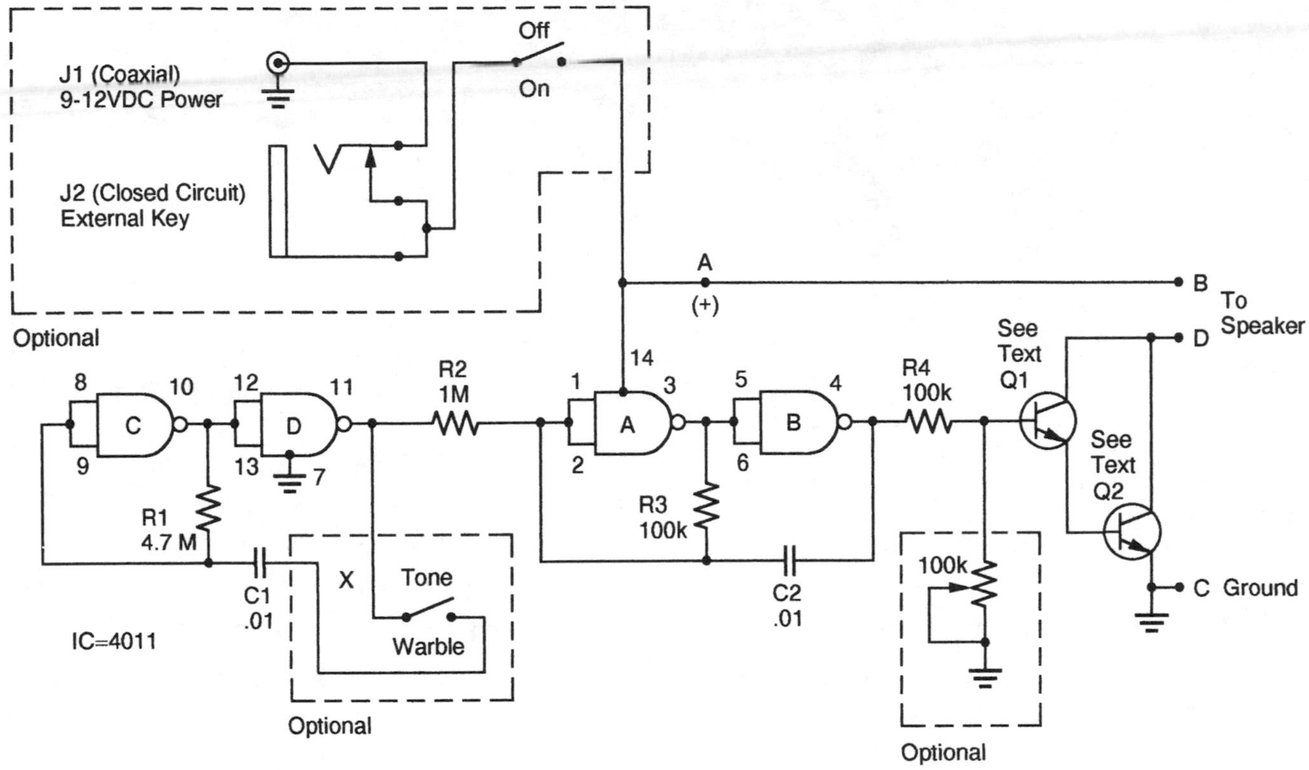

Figure 6-7: Schematic diagram for the "Mad Blaster" siren.

Notice that both inputs of each IC section are wired together, so each section acts just like an inverter: whatever the voltage state at the input pins, the output is the opposite. If the input pins of a section are low, then the output pin of that section is high. Conversely, if the input is high, the output is low. One square wave oscillator is formed with IC sections C and D, together with resistor R1 and capacitor C1. (Normally, without an option to be discussed later, C1 is connected to output pin 11.)

To illustrate circuit action, assume that at a particular instant pins 8 and 9 of IC section C go low. Therefore, pins 10, 12 and 13 go high and output pin 11 of section D goes low. At the instant pins 10, 12 and 13 go high, capacitor C1 begins charging through resistor R1, increasing the voltage at the intersection of R1 and C1, which is also connected directly back to IC input pins 8 and 9. This voltage builds until it reaches "transfer" voltage—the voltage at which change of state to high occurs. This snaps pins 8 and 9 high, forcing pins 10, 12, and 13 to low, and output pin 11 high.

But now capacitor C1 starts discharging through R1 until the R1/C1 intersection voltage drops below threshold—the voltage at which change of state to LO occurs. This snaps pins 8 and 9 low again, back to where we started! This keeps repeating at a rate determined by the values of R1 and C1.

Similarly, sections A and B of the IC together with R3 and C2 form a second oscillator. However, since R3 is a much smaller value than R1, this oscillator changes state more quickly, so it runs at a higher frequency.

Resistor R2 is a means of coupling the output of the section C/D oscillator to force the input section A of the A/B oscillator to follow the frequency of the slow C/D oscillator, thus producing a two-tone effect. If C1 is disconnected (one of the options, shown at point X), the C/D oscillator stops, and the "mad blaster" produces only a single tone.

Resistor R4 couples the warbling or steady tone from IC pin 4 to the base of a pair of NPN transistors wired in a Darlington amplifier circuit. Q1 can be any medium power transistor, but Q2 should be a power transistor. More on this later.

Construction and Modifications

You can build the basic circuitry using perforated phenolic board or make your own printed circuit board as shown in Figure 6-8. If you use the PC board, Figure 6-9 shows the parts layout. Notice that points connecting to external circuitry are shown as A, B, C, and D. Point A is the positive power input, Point C is ground, while Points B and D go to any 8 to 40 Ω speaker or horn. (The parts kit described later includes a drilled and silkscreened printed circuit board.)

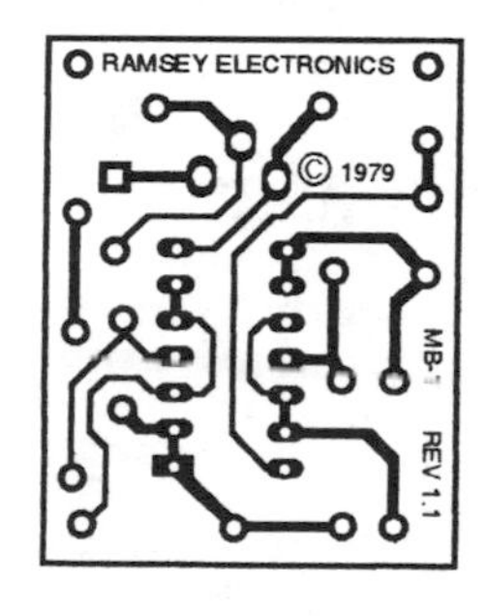

Figure 6-8: Circuit board layout for the "Mad Blaster" siren.

Three options are shown in Figure 6-7 within dotted-line areas. Instead of just directly wiring 9–12VDC power to Point A in the schematic, I elected to use a "Power" jack and a "Key" jack together with a power switch. When wired as shown, with no external key plugged in, power comes in through J1, flows through the closed contacts of J2, and is controlled by the ON/OFF switch.

However, if an external switch, such as a typical code practice telegraph key, is plugged into J2, the upper contacts of the jack separate and current only flows when the external switch is closed. The power switch must, of course, be closed. This allows the unit to be activated from a remote point.

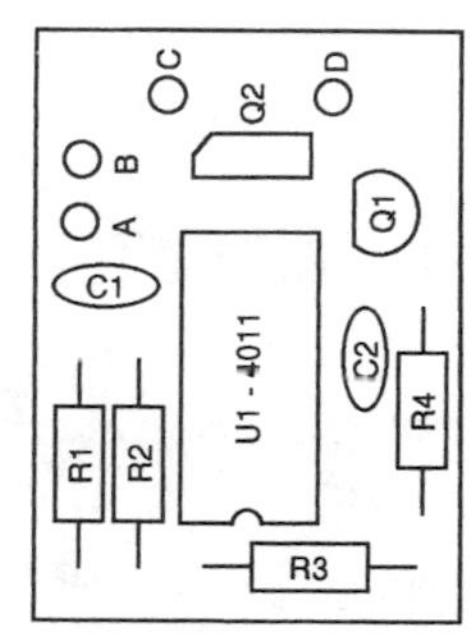

Figure 6-9: Parts placement for the "Mad Blaster" siren.

The "mad blaster" is also capable of single tone operation. By simply using a switch to open the connection of C1 to IC pin 11, this oscillator quits and only the other (high frequency) oscillator is fed to the speaker. The third option involves adding a 50 K or 100 K potentiometer from the base of Q1 to ground. This varies the amount of signal amplified by Q2, therefore acting as a volume control.

Before starting construction, make sure you have all the parts for the basic module and any options you plan. You'll need to have some idea of how you intend to package the unit, and this could determine what kind of connectors you use for the external speaker and power supply. (If you order the parts kit, you'll get the printed circuit board and all the basic parts but you'll have to add your own switches and jacks.)

If you are building from scratch, you must be sure you have NPN transistors for Q1 and Q2. You must also be sure Q1 is a general purpose silicon transistor of the 2N3904 or 2N2222 variety. Transistor Q2 must be a silicon power Darlington type with a gain of about 2000. You must also be sure of the lead identification of both transistors, since this circuit (as shown in Figure 6-6) requires the collectors of both transistors be joined together, and that the emitter of Q1 is directly connected to the base of Q2.

The power output of Q2 generates considerable heat if operation is prolonged. Unless your application only calls for very brief bursts of sound, some sort of heat sinking for Q2 may be needed. Typically, this transistor's mounting hole makes it easy to attach a small piece of aluminum or copper—but be aware that the exposed metal part of the transistor is common to the collector, so you don't want to neatly attach your heat sink to the ground side of the circuit!

Once you've decided what parts and options you'll use, and how you will package the project, the construction is relatively straightforward. I recommend you use a socket for the IC, and make sure pin 1 (the notch end) is properly oriented and seated in the socket. Be very careful that the transistors are not installed "backwards." (Note: This may be different from Figure 6-8 if you don't use the kit transistors!)

Packaging

I recently found an ideal "cabinet" for small electronic projects. It came with some 35mm color slides I had developed, and the local photo shop told me it was a "Fuji Slide Box." It is black, has a hinged snap-down cover, is made from thin plastic, and is rectangular in shape. The outside dimensions are 4.6x2.2x1.2 inches.

Inside, with the exception of four 0.25-inch ridges that extend inside the slide box near its center, there is plenty of room for a small circuit board and one or two 9-volt transistor radio batteries. Since the plastic is so thin, it is easily cut with a sharp knife or drill to allow placement of jacks and switches. Figure 6-10 shows how I managed to get the PC board, the volume control potentiometer, all switches, and connecting wiring inside one of these slide boxes.

Figure 6-10: Interior view of the "Mad Blaster" housed in a Fuji slide box.

Where do you get these boxes? I had the photo shop order some for me at 35¢ each! Also, processed slides often come packed in these boxes. If you have some friends who are into slide photography, they may have a spare box or two.

Testing

The speaker is connected between points B and D as shown in Figure 6-7. When you select a speaker to use with the "mad blaster," be sure it can handle the maximum allowable power, about 15 watts. Typically, this will be a public address or paging horn-type speaker, or an automobile horn.

If you use the optional volume control you can, of course, use a smaller speaker, such as a standard 4-inch or larger utility speaker. The speaker impedance should ideally be 8 Ω , although 4 to 40 Ω speakers should also work.

For power, a 12-volt automobile battery will do well. If you're using a power supply, make sure it can deliver very low-ripple 12 volts DC up to 1 ampere. Most wall-plug transformers have too much ripple and too low a current capability, so they are not adequate for this application. If you wish to use batteries, use eight C or D cells in series.

Connect the speaker and power and hold your ears! If your "mad blaster" has a volume control, be sure it is not turned below the audible range. If you don't hear anything after turning up the volume, use a voltmeter to check the power voltage.

Troubleshooting

Still does not work? Be sure the integrated circuit is not installed backwards, and that the two transistors are wired together as shown in the schematic.

If the 4011 IC is working properly, you can hear it through a separate test amplifier connected to the base of Q1. The sound should be stronger at the base of Q2. If it isn't, Q1 is suspect. If it is stronger at the base of Q2 but the speaker is still silent, then Q2 is the problem.

At the time this book was being written, a kit which included all the parts marked with an asterisk (*) above was available from Ramsey Electronics, Inc., 793 Canning Parkway, Victor, NY, 14564; phone 1-800-446-2295. Their stock number was MB-1. The cost per kit was approximately $5, although there was a shipping charge of $4.95 per order and an additional $3 charge on all orders under $20. Availability, pricing, and ordering policies may have changed since publication of this book.

MAD BLASTER PARTS LIST

* C1, C2:	0.01 µF disk capacitor
* IC:	4011 CMOS quad 2-input NAND gate
* R1:	4.7 megohm 1/4 W carbon resistor
* R2:	1 megohm 1/4 W carbon resistor
* R3, R4:	100 kilohm 1/4 W carbon resistor
* Q1:	2N3904 NPN silicon transistor or equivalent (see text)
* Q2:	SK-3996 or NTE253 or ECG253 NPN Darlington power amplifier transistor or equivalent (see text)
Miscellaneous:	
	Battery or power supply, speaker capable of handling up to 15 watts.
Optional:	* Printed circuit board, 100 kilohm volume control, power switch, warble/tone switch, key jack, 14-pin IC socket, case, Q2 heat sink, external switch or telegraph-type key.

CHAPTER 7

A Variable Frequency Audio Oscillator

This circuit was designed as an audio oscillator for learning the international Morse code. Thus, we'll call this project "Code Buddy." However, it is also a versatile variable frequency audio oscillator. This circuit is built around the common, and inexpensive, 555 timer IC and a few additional parts.

For years, anyone wanting to get a ham radio license had to be able to send and receive messages in Morse code. As a result, "code practice oscillators" (CPOs) were popular construction projects for would-be hams. These CPOs were actually sine wave audio oscillators, usually with a variable output frequency, and many non-ham electronics experimenters were able to adapt these circuits for their own purposes.

Today, the most popular class of ham license issued by the Federal Communications Commission, the technician class, doesn't require a Morse code test (a written exam on electronics and radio is required, however). However, international treaties still require a Morse code test for ham operators who have access to shortwave frequencies capable of worldwide communication. This means CPO circuits are still common, except today they are built around integrated circuits instead of transistors or vacuum tubes. This chapter describes such a circuit. The "Code Buddy" is useful whether you're studying for your ham radio license or just need a wide frequency range audio oscillator for your electronics workbench.

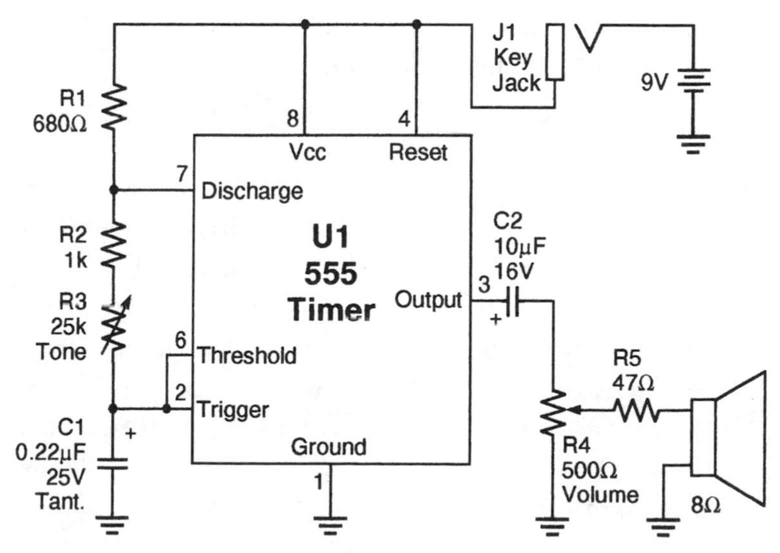

Figure 7-1: Schematic of the "Code Buddy" audio oscillator.

Circuit Description

Figure 7-1 shows the schematic of the Code Buddy. As you can quickly see, the heart of the design is the popular 555 IC timer. Produced by several different manufacturers (who vary the internal design somewhat), the most common version of this IC has 23 transistors, two diodes and 16 resistors on the silicon chip within its package

Figure 7-2 shows a block diagram of a typical 555 timer. Its components form two comparators, a flip-flop, an output stage, a reset transistor and a discharge transistor. The supply voltage, Vcc, is applied to Pin 8, and Pin 1 is GROUND. Pin 2 is connected to an input to the TRIGGER comparator, and Pin 6 is connected to an input of the THRESHOLD comparator.

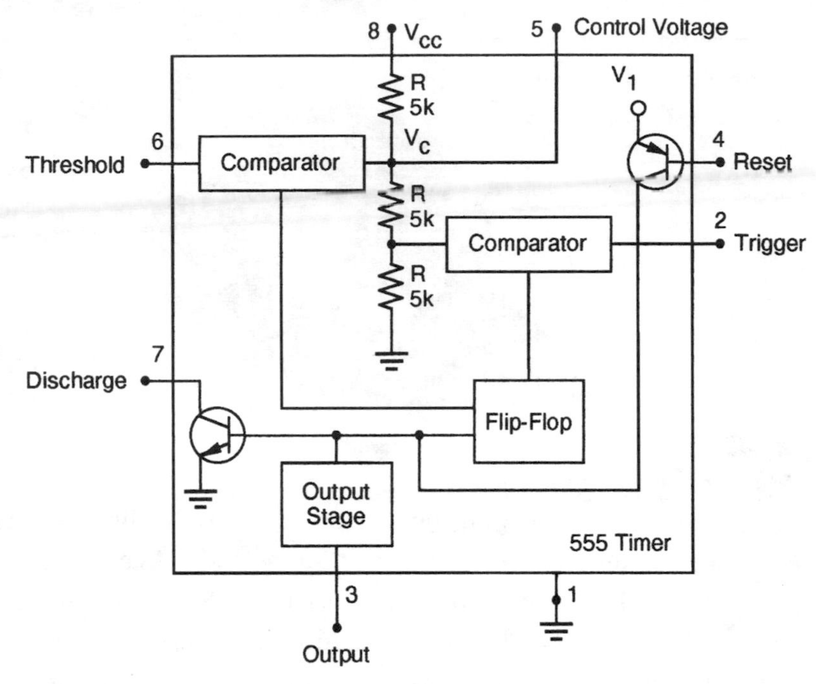

Figure 7-2: Internal diagram of 555 timer IC.

Notice that three resistors of equal value are placed between Vcc and GROUND so that 1/3 Vcc appears as a reference voltage to an input of the TRIGGER comparator, and 2/3 Vcc is the reference voltage at an input of the THRESHOLD comparator. Since the resistors are all of the same value, the 1/3 Vcc and 2/3 Vcc points are relative to Vcc, and therefore independent of actual Vcc variation.

The OUTPUT (either high or low) appears at Pin 3, while a DISCHARGE transistor provides a path to GROUND at Pin 7 whenever the OUTPUT is low. When the RESET transistor at Pin 4 is turned ON with a voltage below 0.4 volts, it makes the OUTPUT at Pin 3 low. When not intended to be used for RESET, Pin 4 is connected to Pin 8 to prevent spurious resetting. The Pin 5 CONTROL VOLTAGE is not used in this application.

The 555 has two principal operating modes: "monostable," which must be triggered to change state; and "astable," which changes state automatically (or "oscillates") depending on the external components used. In this circuit, the 555 is configured for use in the astable mode as a simple variable frequency audio oscillator with volume control.

When a key is plugged into the J1 key jack, it acts as a switch. If you don't want to use this circuit for Morse code practice, you can substitute a SPST switch to turn the tone on and off. (A pushbutton SPST switch is a handy way to produce "bursts" of tone.) When the key is closed, battery voltage, Vcc, appears at Pin 8 and Pin 4 to power the IC and prevent spurious resetting. Since capacitor C1 is uncharged, the voltage at Pin 2 is below 1/3 Vcc and therefore the OUTPUT at Pin 3 is high. Voltage begins to build up on capacitor C1 through resistors R1 and R2, and potentiometer R3. When the voltage on C1 reaches the THRESHOLD voltage (2/3 Vcc at Pin 6), the OUTPUT at Pin 3 goes low and the DISCHARGE transistor at Pin 7 provides a path to GROUND. Now C1 starts to discharge at Pin 7 at a rate depending on the value of resistor R2 and the setting of potentiometer R3. When the voltage at C1 drops below the TRIGGER voltage (1/3 Vcc at Pin 2), the OUTPUT goes high and the DISCHARGE path at Pin 7 is opened. C1 starts to charge again.

Oscillation occurs because the TRIGGER and THRESHOLD inputs are both tied to the positive side of capacitor C1, which is continually charging and discharging between 1/3 Vcc and 2/3 Vcc at a rate controlled by R1, R2, and the setting of tone potentiometer R3. Capacitor C2 couples the square-wave output of U1 to potentiometer R4, which controls the signal volume through resistor R5 to the small speaker.

Construction

The Code Buddy circuit uses commonly available parts you can get from most electronics suppliers. In addition, a parts kit (including a printed circuit board) was available when this book was being written; it will be described later.

While you can use perfboard construction (or even prototype the circuit with a solderless breadboard), construction is greatly simplified if you use a printed circuit board. Figure 7-3 shows the printed circuit board, foil side, in actual size. Figure 7-4 shows the component layout using this board.

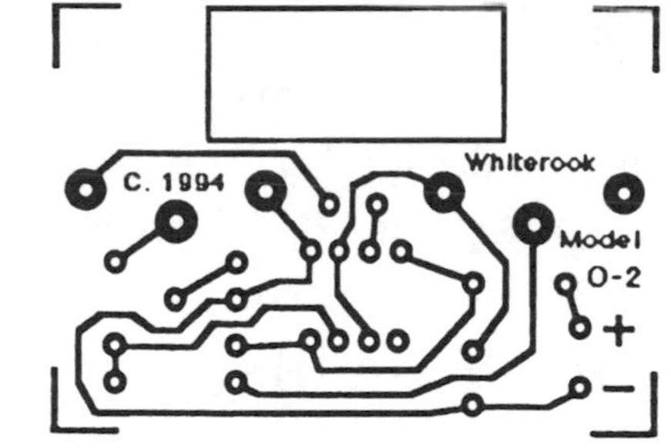

Figure 7-3: Foil side of printed circuit board.

Use only rosin core solder and a 25–35 watt pencil type soldering iron. When installing parts into the printed circuit (PC) board, be sure that capacitors C1 and C2 are installed with the positive leads oriented as shown, and that the U1 integrated circuit is plugged into its socket with the notch (Pin 1 end) as shown. C2 will have to be installed flat against the board so it does not interfere with the front panel in later assembly. Similarly, the potentiometers will also have to lie flat against the board. When soldering the battery snap to the PC board, make sure the red wire is soldered to the plus (+) pad. Make sure the key jack you are using is an open circuit type.

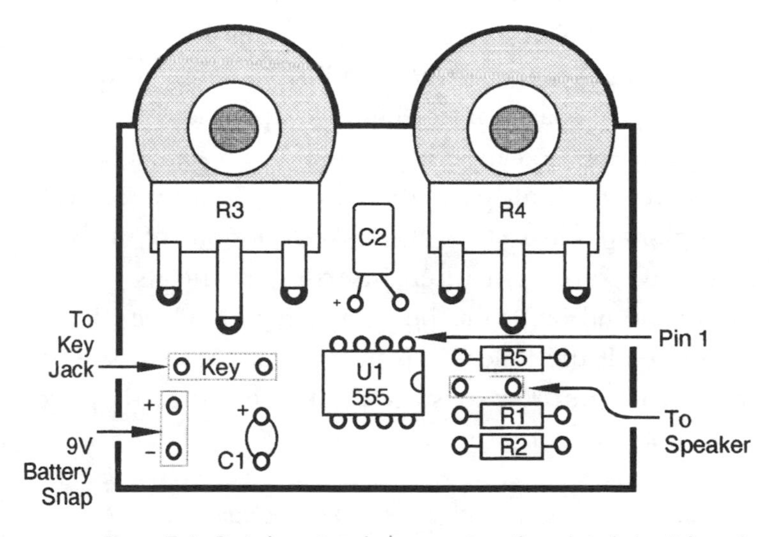

Figure 7-4: Parts layout and placement on the printed circuit board.

The final packaging and assembly (placement of the speaker, potentiometers, key jack, etc.) can be done in a variety of ways. (The parts kit includes a pre-drilled cabinet and labeled front panel, which greatly simplifies things.) Figure 7-5 shows the assembled Code Buddy circuit, built from the parts kit, prior to being placed in a cabinet. A piece of double-sided foam tape holds the speaker to the foil side of the PC board in this illustration. You'll need to connect short hookup wires from the PC board to the key jack and the speaker terminals—there's no polarity consideration in either case. You'll also need to drill holes in the cabinet for the speaker to be heard if you're not using the parts kit.

Figure 7-5: The assembled "Code Buddy" before being placed in its housing.

Testing

For best results, always use an alkaline 9-volt battery. If you use a closed-circuit key jack by mistake, the Code Buddy will sound off when you connect the battery. This should not be! Power should not be applied to the circuit until an EXTERNAL key is closed. No power is drawn from the battery until the external key is closed.

To test the unit, simply connect an external key through a two-wire cable (with the appropriate plugs on each end) to the key jack on the Code Buddy front panel. When the external key is closed, you should hear a tone through the speaker. The frequency is controlled by the TONE control, the loudness by the VOLUME control.

Troubleshooting

If Code Buddy does not work properly (or at all!) be aware that the majority of problems arise from construction errors. The most common problem, by far, comes from solder bridges between traces and pads on the PC board. Hold the PC board up to a light, compare it with the PC board drawing, and any bridges should be readily apparent.

Next, check all parts that involve polarity: C1, C2, battery snap, Pin 1 of the IC. Make sure none of the pins of the IC were bent when inserting it into its socket. There should be 9 volts between Pin 8 (Vcc) and Pin 1 (ground.) If the tone seems unstable or scratchy sounding, clean your key or switch contacts, since they are likely corroded or dirty.

NOTE: *A complete kit of all parts above, including cabinet and labeled front panel, was available at the time this book was being written from Whiterook Products Company, 309 S. Brookshire Ave., Ventura, CA 93003; phone 805-339-0702. The price was $21.95 and shipping was $6.50. An assembled version was also available. Availability and pricing may have changed since publication of this book.*

PARTS LIST	
C1:	0.22 μF (10 WVDC or higher) tantalum capacitor
C2:	10 μF (10 WVDC or higher) electrolytic capacitor
R1:	680 Ω 1/4 W 5% carbon resistor
R2:	1 kilohm 1/4 W 5% carbon resistor
R3:	25 kilohm potentiometer with mounting hardware and knob
R4:	500 Ω potentiometer with mounting hardware and knob
R5:	47 Ω 1/4 W 5% carbon resistor
J1:	3.5mm (1/8-inch) miniature open-circuit phone jack
U1:	555 timer integrated circuit
Speaker:	8 Ω 2-inch diameter
Snap:	9 volt battery snap connector
Socket:	8-pin IC socket for U1
Miscellaneous:	printed circuit board, double-sided tape, hookup wire, solder.

CHAPTER 8

Building a VOX (Voice-Operated Switch)

A voice-operated switch (commonly called a VOX) is a switch that is "on" so long as a voice or other sound is louder than a certain level. The VOX described in this chapter can be used to voice-activate a tape recorder, a radio transmitter, an alarm, etc.

Sound-activated devices are not new. For years a device has been available that allows you to turn your television "on" by clapping your hands, and then turn it "off" by clapping your hands again. This same device could, of course, also be used to turn a lamp or motor on and off. But its basic operation consists of latching a circuit on or off, and waiting for the next "command." If it is on, the next command (clapping hands, shouting, or any other loud noise) will shut it off. If it is off, the next command will turn it on. This is sometimes called *toggling*, where each action causes a change of state.

A "VOX" (voice-operated switch) circuit, on the other hand, takes only a voice (or other sound) to trigger it "on." When triggered, it stays on ONLY as long as the sound remains above a certain (usually adjustable) threshold volume. When the volume drops below this threshold, the circuit turns "off" after a short delay.

VOX circuits have been used for many years in radio communications, especially in situations where the radio operator may not have a free hand to hold a microphone or press its "push-to-talk" switch. Since most radio communication is either in the send or receive mode, it's convenient for the radio to switch to "transmit" when the operator starts speaking, and back to "receive" soon after the operator stops speaking. This also encourages a more natural, conversational way of speaking over the radio.

VOX circuits are also widely used with tape recorders, since they can be used to turn the recorder tape transport motor on whenever a person starts speaking. This saves a lot of tape when the user is thinking instead of speaking. It also allows placing a recorder in a hidden location for clandestine recording when someone begins to speak.

Typically, a VOX supplies voltage to close a relay when a certain sound level is exceeded. What you operate with the relay contacts is up to you. Any circuit that you would like to trigger during the period when someone is speaking louder than a certain level is a good candidate for use with the VOX circuit I'll describe in this chapter.

Circuit Description

Figure 8-1 shows the schematic for a VOX you can build from commonly available parts, or a parts kit, including printed circuit board, described later in this chapter. This circuit runs on from 6 to 12 volts DC, and will drive up to a 100 milliampere (mA) load, such as a tape recorder motor or relay.

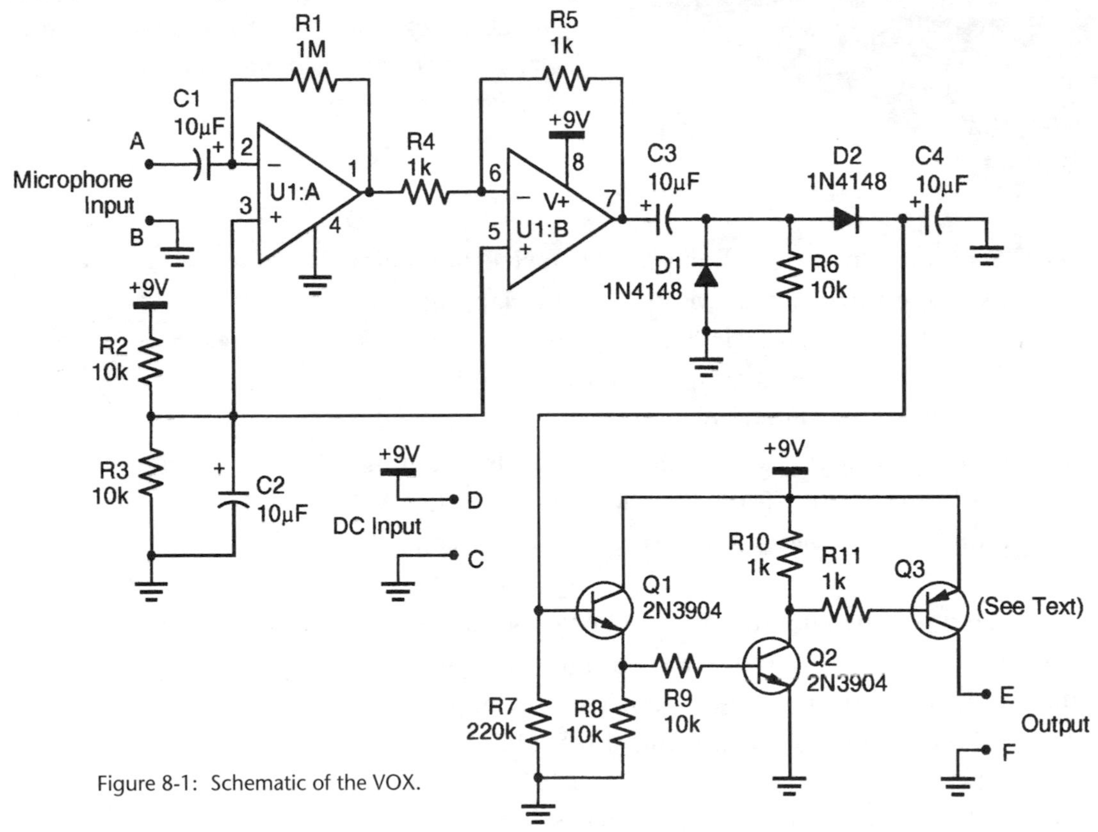

Figure 8-1: Schematic of the VOX.

The microphone input at Points A and B is amplified by the first section of U1, an LM358 dual operational amplifier integrated circuit. Electrolytic capacitor C1 blocks DC from the microphone, yet passes the alternating-current sound signal to the Pin 2 negative input of U1:A. Resistors R2 and R3, and electrolytic capacitor C2, configure the positive supply voltage to Pins 3 and 5 to allow the LM358 to operate from a single power supply. (Many operational amplifiers require a dual power supply.)

The Pin 1 output of U1:A is fed back to the input of U1:A by resistor R1, and also passed on to U1:B Pin 6 by resistor R4. With feedback resistor R5, the microphone signal is further amplified and fed from Pin 7 to electrolytic capacitor C3. The signal at this point is still alternating current, so it passes through C3 to diodes D1 and D2, and resistor R6. These rectify the amplified microphone signal to a direct current level, with electrolytic capacitor C4 smoothing out the voltage level and acting as part of a resistor-capacitor (RC) timing circuit together with resistor R7. The voltage level at this point operates the switching circuit composed of transistors Q1, Q2, and Q3 and their associated resistors.

The voltage at the top of R7 provides enough positive base bias to put NPN transistor Q1 into full conduction, with collector-to-emitter current supplied directly from the power source. When Q1 is conducting, the voltage at its emitter-resistor R8 junction provides enough positive base bias through resistor R9 to put NPN transistor Q2 into full conduction, with collector-to-emitter current provided through resistor R10 from the power source.

When Q2 is fully conducting, the junction of resistors R10 and R11 is essentially at ground (negative) potential, which provides enough negative base bias to PNP transistor Q3 to switch it into full conduction. This places Point E at essentially supply voltage compared to Point F. Switched emitter-to-collector current through Q3 can therefore be used to directly drive a device (up to about 100 mA) between Points E and F, or to operate a relay between these points for greater current requirements.

As long as incoming sound through the microphone keeps the voltage at the top of R7 high enough to bias Q1 into full conduction, the output voltage remains between Points E and F. However, when sound at the microphone drops to a level below the threshold required, the voltage at the top of R7 and on the positive side of C4 starts bleeding off through R7 to ground. The values for R7 and C4 provide about a 1-second RC time delay after "silence" at the microphone before the switching circuit (Q1, Q2, Q3) shuts off Q3, removing voltage at Point E.

Construction

Figure 8-2 shows the layout of the printed circuit board for this circuit (an etched and drilled board is included in the parts kit). Figure 8-3 shows the correct parts placement if you use the PC board. If you prefer, a perforated board could be used for the construction. However, a PC board is a better approach since circuits built on a perforated board could suffer from stray oscillations that result in false "on" and "off" states for your VOX.

Most of the parts are not critical, but you may want a medium-power switching transistor for Q3 if you intend to

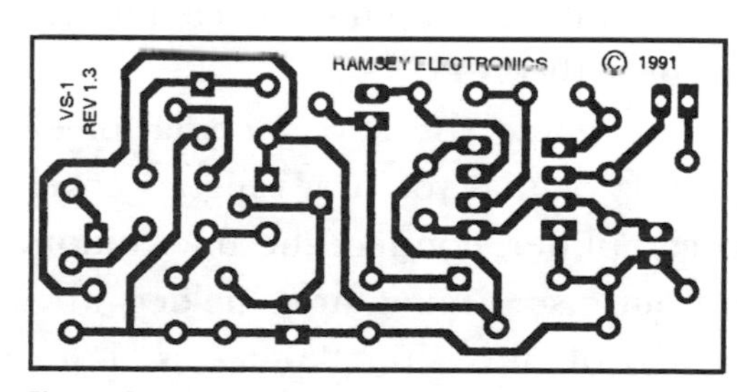

Figure 8-2: Printed circuit board layout.

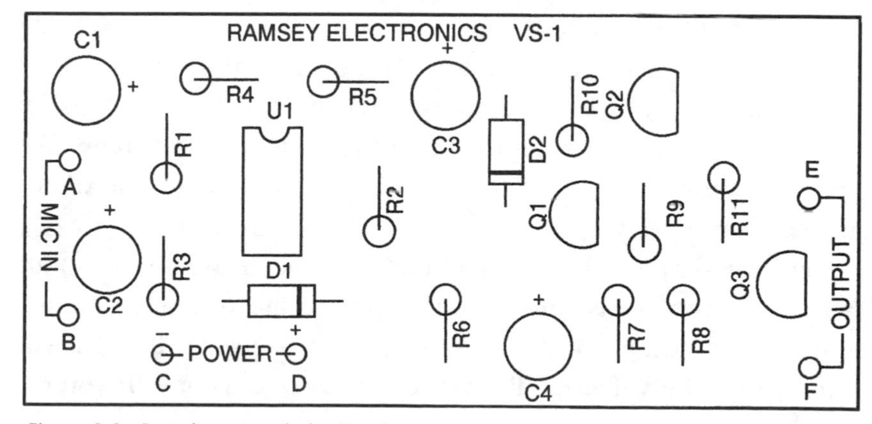

Figure 8-3: Parts layout and placement.

have it conduct more than 100 mA. The transistor included in the parts kit is a "228256," which is some sort of house number (probably similar to a 2N3906), apparently rated for 100-mA maximum emitter-collector current. But if you use an MPS2907 PNP switching transistor (available from Radio Shack at the time this book was written as their catalog number 276-2023) it will do nicely up to 600 mA, and comes in the same TO-92 case and lead configuration.

Be sure the diodes and capacitors are oriented with the polarity as shown in Figure 8-3, and the transistor flat sides are in the direction shown. Also be certain the integrated circuit has the end notch or dot placed as shown. Since many of the resistors have similar values (10K or 1K), be careful not to mix them up.

Testing the VOX

Although the manual that accompanies the parts kit states that the completed VOX can be used "with any type of microphone," I did not find this to be the case. Crystal or electret condenser microphones work great, since they have relatively high output. A source for an electret microphone with built-in preamplifier is included in the parts list.

However, I tested my VOX with various dynamic low-impedance microphones—the kind commonly used as external microphones with tape recorders—and they did not produce enough voltage to operate the VOX without a preamplifier. Figure 8-4 shows the circuit of a simple one-transistor preamplifier I built using an NPN 2N3904 transistor. This can be built on a piece of perforated board, with its output connected to the microphone input of the VOX.

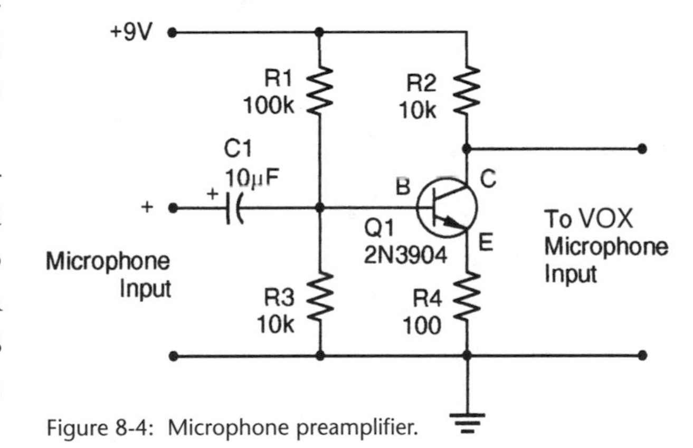

Figure 8-4: Microphone preamplifier.

Assuming you have a crystal or electret microphone, or that you are using a weaker mike with a preamplifier, connect the microphone or preamp output, using miniature shielded wire, to Points A and B of the VOX. Point A is "hot" and Point B should be connected to the shield (grounded side) of the microphone cable.

The VOX circuit is designed for optimum operation at 9 volts DC, but you may go up to 12 volts. The addition of a 7808 3-terminal voltage regulator will provide an 8-volt regulated voltage. For satisfactory 6-volt operation, you'll need to increase the value of R7 for an acceptable delay.

For the simplest first test, connect a voltmeter to Point E (+) and Point F (–). When you speak into the microphone the voltmeter should instantly jump to a reading a little below the power supply voltage. If you make a brief sound the voltage will hold for about one second. If you continue to speak steadily, the output will remain powered without interruption, and drop out about one second after you stop speaking. You can adjust the delay time, as discussed later.

Troubleshooting

If the VOX does not appear to be working, check the orientation of all parts, as described previously and shown in Figure 8-3, and be sure the proper resistor values are in the proper location. Of course, look at all solder joints for any that are not bright and shiny. If you have an oscilloscope, you can check from the microphone input on through the circuit for a point where the microphone signal drops out. The greatest amplified microphone signal should be between Pin 7 of IC1:B and ground. Beyond that point you'll have a DC level created by the microphone signal.

Using the VOX

The most practical output device to connect to Points E and F is a small low-current 6 or 12 volt DC relay, such as any of several available from Radio Shack and other sources. The contacts of the relay should be rated for the device to be switched.

Many tape recorders have a "remote" jack, which simply connects to an external switch, usually on the microphone. When the recorder amplifier is on, but the switch is open, the motor is off and the tape is stationary. When the external switch is closed, the motor turns on, transporting the tape for a recording.

If you intend to control a tape recorder motor with the VOX, first measure the motor's current requirements with a multimeter by putting the multimeter milliampere function in series with an external switch. If the motor uses less than 100 mA, you can use the E-F terminals of the VOX (Q3's switched output) to control the motor directly. If the recorder motor requires more than 100 mA (or more than 600 mA if you use an MPS2907 for Q3), then use a relay.

If using the VOX with a transmitter, the output of the VOX (using a relay) is used to operate the PTT (push-to-talk) relay in the transmitter, usually operated by a microphone pushbutton switch. See Figure 8-5. If you don't want to parallel the transmitter mike to the VOX, use a separate electret microphone for the VOX input.

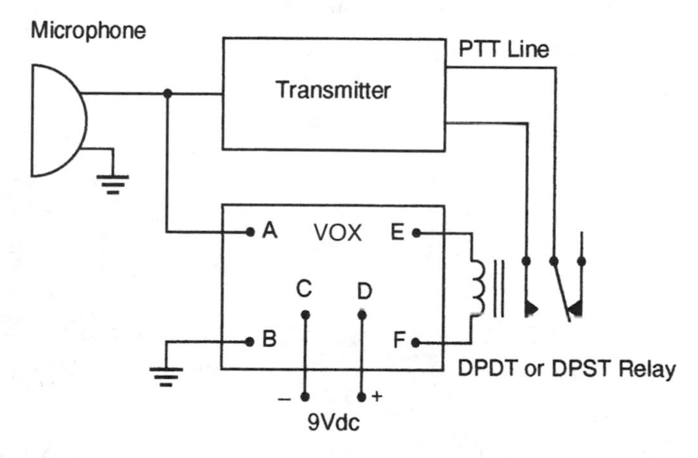

Figure 8-5: Connections for using the VOX with a radio transmitter.

Modifications

The VOX delay time before dropout of the voltage at Points E and F is controlled by the values of C4 and R7. A 1 megohm trimmer potentiometer in place of R7 will provide an excellent delay adjustment. Alternately, you can parallel another electrolytic capacitor—about 2 to 5 μF—across C4 to extend the delay time.

Most commercial VOXs contain "anti-trip" or "anti-VOX" circuitry to prevent signals or noise coming from the speaker from tripping the VOX. Effective anti-VOX requires additional differential amplifier or comparator circuitry that is beyond the purpose of this design. There are several things you can do to minimize unwanted tripping: You can control the sensitivity of your VOX by using a 2 K potentiometer at the microphone input, as shown in Figure 8-6; you can keep the receiver speaker volume low; you can increase the distance between the speaker and the VOX microphone.

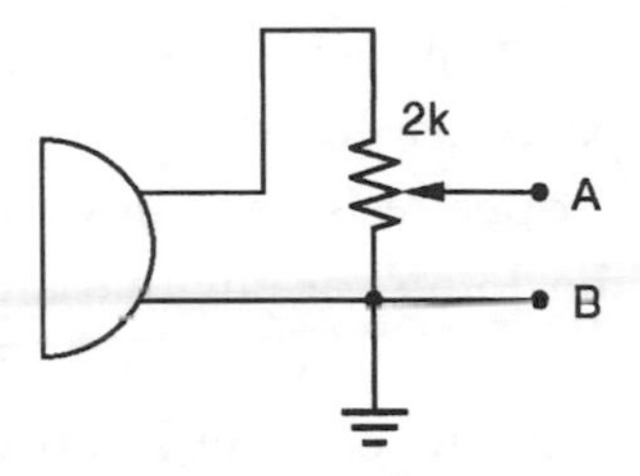

Figure 8-6: Optional VOX sensitivity control.

Packaging

Regardless of whether you build this VOX circuit "from scratch" or using the parts kit, the complete circuit will have input, output, and battery wires hanging free. If you use the optional preamplifier, you will also have wires going to the mike input, preamplifier output, and battery.

Some sort of housing or packaging is clearly needed. I assembled the VOX and the preamplifier in a "Fuji Slide Box," as previously described in Chapter 6.

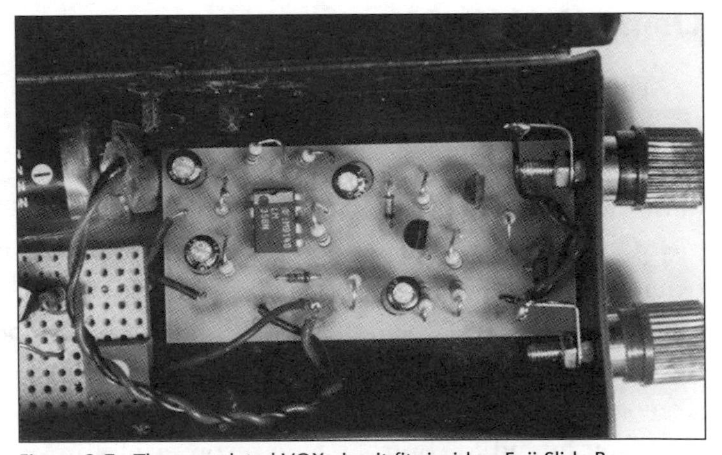
Figure 8-7: The completed VOX circuit fits inside a Fuji Slide Box.

The Fuji Slide Box is just the right size for the VOX PC board, the preamplifier, and a standard 9-volt battery. A microphone jack is used at the input, and binding posts used at the output. The battery may be used to parallel power the VOX and the preamplifier, being careful to observe battery polarity for both. Figure 8-7 shows how I managed to fit everything into a single Fuji Slide Box. Figure 8-8 shows the neat, compact unit you get when this packaging is used.

Figure 8-8: The Fuji Slide Box makes a convenient housing for the circuit.

A complete kit of parts, including an etched and drilled printed circuit board and a detailed assembly/instruction manual (but not the Fuji Slide Box), was available at the time this book was being written from Ramsey Electronics, Inc., 793 Canning Parkway, Victor, NY 14564; phone 1-800-446-2295. Their stock number was VS-1 and the cost was $6.95 per kit. An MC-1 miniature electret condenser microphone was also available for $2.95. There was also a shipping charge of $5.95 per order with an additional $3 charge for orders under $20. Availability, pricing, and ordering policies may have changed since publication of this book.

VOX PARTS LIST

C1, C2, C3, C4:	10μF 15WV electrolytic capacitor
D1, D2:	1N4148 or 1N914 silicon signal diode
Q1, Q2:	2N3904 NPN general purpose transistor
Q3:	PNP medium power transistor (see text)
R1:	1 megohm 1/4 W 5% carbon resistor
R2, R3, R6, R8, R9:	10 kilohm 1/4 W 5% carbon resistor
R4, R5, R10, R11:	1 kilohm ohm 1/4 W 5% carbon resistor
R7:	220K kilohm 1/4 W 5% carbon resistor
U1:	LM358 8-pin DIP dual operational amplifier
Optional:	Fuji Slide Box or other packaging

OPTIONAL PREAMPLIFIER PARTS LIST

C1:	10 μF 15 WV electrolytic capacitor
R1:	100 kilohm ohm 1/4 W 5% carbon resistor
R2, R3:	10 kilohm ohm 1/4 W 5% carbon resistor
R4:	100 Ω 1/4 W 5% carbon resistor
Q1:	2N3904 NPN general purpose transistor

CHAPTER 9

A Nickel-Cadmium Battery Zapper

A lot of nickel-cadmium ("NiCad") batteries are thrown away needlessly every year. Many of those "defective" NiCad batteries could be restored to normal operation by using a device like the "NiCad Battery Zapper" described in this chapter. You may never have to throw away a NiCad battery again!

Nickel-cadmium (NiCad) rechargeable batteries are almost as ubiquitous as bacteria and viruses! They're used in wireless and cellular phones, computers, calculators, flashlights, portable radios and televisions, camcorders, rechargeable power tools, and toys of all sorts. And while they normally can be recharged and used again many times, sometimes NiCads refuse to charge and are thrown away.

NiCad batteries have less operating capacity for their size than non-rechargeable zinc-carbon and alkaline cells. However, NiCads can be recharged and used up to 1000 times, thus making them the least expensive in the long run. They also can provide higher current, which is required in some applications. The downside of NiCads, in addition to their greater original cost, is the bother of recharging them, and the frustration when they "die" virtually without warning.

What Are the Problems?

The chemistry of nickel-cadmium batteries is complex, and the details are well beyond the intent of this chapter. Suffice to say that nickel-cadmium batteries are pretty particular about their operating environment and how they are discharged and recharged.

Their normal charged voltage is 1.2 volts, and they will operate with a relatively flat discharge curve to about 1.1 volts, at which point they begin to drop voltage rapidly. The usable period of operation above 1.1 volts depends, of course, on the discharge rate and the original battery charge.

So-called "memory" problems, wherein the NiCad battery seems to "remember" its last operating voltage before recharging, and virtually quits at that voltage in successive uses, can be defeated by what is called a "deep discharge." Properly, a NiCad cell should be discharged through a resistance at a current of about one-third of its rated hourly life (say, 50 milliamperes for a 150 milliamp-hour (mah) rating) until the voltage is no less than 0.5 volts per cell. Unfortunately, if you discharge below that voltage, you may damage the cell.

In the case of NiCads in series, such as battery packs of 2.4, 3.6, 6, or 12 volts used in phones and computers, the weakest cell can reverse polarity and make the whole pack unusable! Although only one cell in the pack may be bad, you are faced with a buying an expensive replacement pack.

There are some potential problems in the recharging process. If you overcharge NiCad cells with too much current for too long—or at too high an ambient temperature—the NiCads will vent the gas caused by a pressure build-up even though the cells are "sealed." This leads to a premature loss of capacity. The solution is to use a current-regulated charger for that cell or pack.

One common problem is that defects in manufacture, abusive use, complete discharging, non-use, and old age allow "whiskers" (sometimes referred to as "dendrites") to form between the positive and negative terminals of a NiCad cell. These tiny, thread-like paths are essentially short circuits that allow the battery to self-discharge. If the whiskers are not excessive, the battery can be charged, but it will self-discharge in a short time. If the whiskers are prolific, the battery will refuse to charge to its normal voltage, or may not accept any charge! That's when the battery seems to be dead and is usually thrown away.

Save That Battery!

This is where our "NiCad Battery Zapper" comes in. This device is designed to blast the inside of the battery between the positive and negative terminals with a burst of current that will vaporize the whiskers and clear the battery back to essentially its original state.

In the case of battery packs where the cells are generally connected in series, you can access each cell to determine which (it's usually just one) is the culprit and then zap it, restoring the whole pack to normal operation. This can save $100 or more for typical laptop computer battery packs!

How do you access a single cell in a pack? If the pack is sealed in plastic shrink-wrap, use straight pins to pierce the shrink-wrap where the bulges indicate battery connections. If the pack is sealed in hard plastic, more drastic measures are needed. You can usually wedge, cut, or saw the pack open. If it is destroyed in the process, remember: it was no good anyhow! (You might be able to salvage the good cells for other purposes, however.)

The NiCad Battery Zapper uses several unusual parts that you might have trouble locating. However, a parts kit is available, as we'll describe later in this chapter.

Circuit Description

The schematic for the NiCad battery Zapper is shown in Figure 9-1. You need to be careful when building and testing this circuit, for it produces a high voltage (up to 180 volts!) that results in several amps of current to "blast away" the whiskers that form inside NiCads.

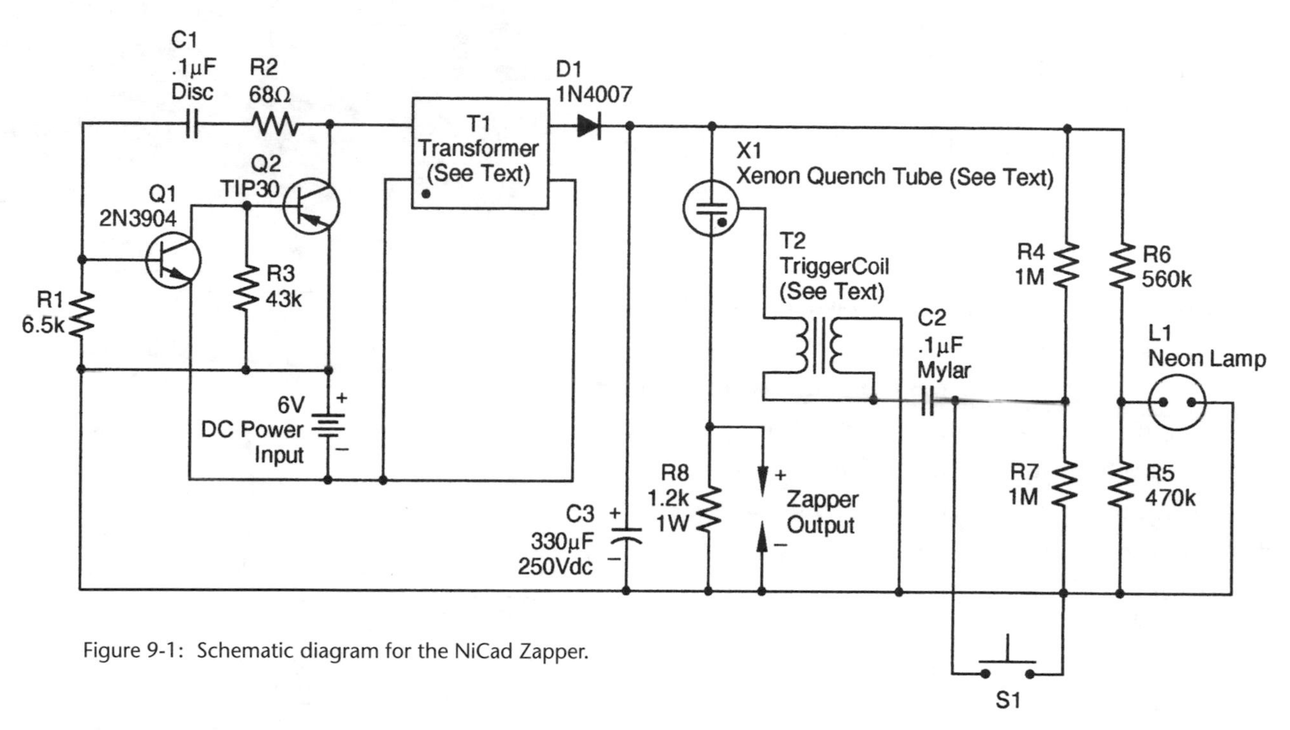

Figure 9-1: Schematic diagram for the NiCad Zapper.

Transistors Q1 and Q2, and associated components R1, R2, R3, and C1, form an oscillator which converts the input DC (typically 6 volts, and shown as a battery pack in Figure 9-1) to an AC voltage that is fed to the primary of transformer T1. This voltage is stepped up by T1's secondary and rectified by diode D1. The current through D1 charges large electrolytic capacitor C3, building up the voltage to approximately 180 over a period of a few minutes.

When enough charge is accumulated across C3 (160 VDC or more), the voltage at the divider formed by the junction of R5 and R6 reaches the firing point for neon lamp L1, lighting it to indicate that the circuit is almost ready for zapping.

During the charging time, the voltage at the junction of resistors R4 and R7 slowly charges capacitor C2. When pushbutton switch S1 is pressed, C2 suddenly discharges through S1 to ground, creating a pulse through the primary of trigger transformer T2 (right side in Figure 9-1).

The secondary of T2 provides a high voltage trigger pulse to the xenon flash ("quench") tube X1. X1 flashes and provides a very low resistance that discharges C3 instantly through the output leads that connect to the NiCad being "zapped." Resistor R8 allows a discharge path if no battery is connected to the output.

Construction

The printed circuit (PC) board pattern for this circuit is shown in Figure 9-2, and the parts layout is shown in Figure 9-3. The parts list at the end of this chapter shows the required components and possible sources.

If you have a good "junk box" of parts, you might want to build this device from scratch. However, several parts—T1, T2, X1, and C3—may be hard to find. If you don't have the exact values and types for these four components specified in the parts list, don't attempt to substitute other components. The circuit may not work properly and you could damage the NiCad cells instead of restoring them.

A drilled and etched PC board is included with the parts kit. If you prefer, you can make your own PC board or use insulated perfboard. If you use perfboard, group the parts in the same fashion as shown in Figure 9-3. Install all the components as shown in Figure 9-3, strictly observing polarities and the facing directions shown for the transistors. For easiest installation of other components, install C3 last.

The xenon quench tube, X1, comes in a plastic bag and its leads have been coated with a special flux for good soldering. Hold X1 by the glass part to keep from running the flux off its leads. It mounts with its dot facing Mylar capacitor C2. Make sure that the leads of X1 are firmly soldered. Also note that the wire extending from the top of T2 is bent over and soldered to the copper foil band around X1.

The red clip lead connects to the (+) point on the output, and the black clip goes to the (–) point. Connect push-button switch S1 to the appropriate points on the circuit board using small lengths of insulated wire. Your completed circuit board should look like the one in Figure 9-4.

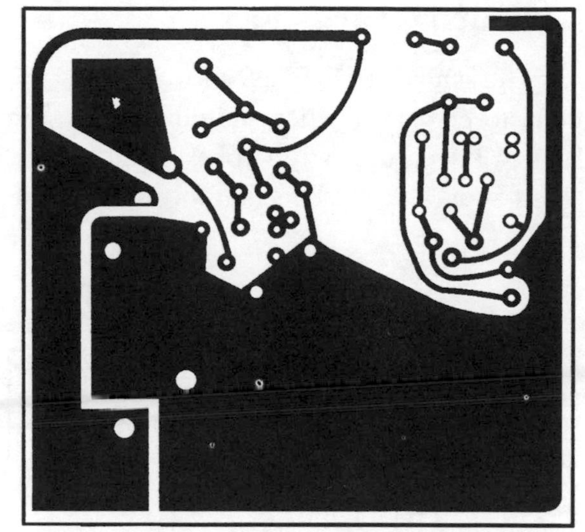

Figure 9-2: Printed circuit board pattern for the Zapper.

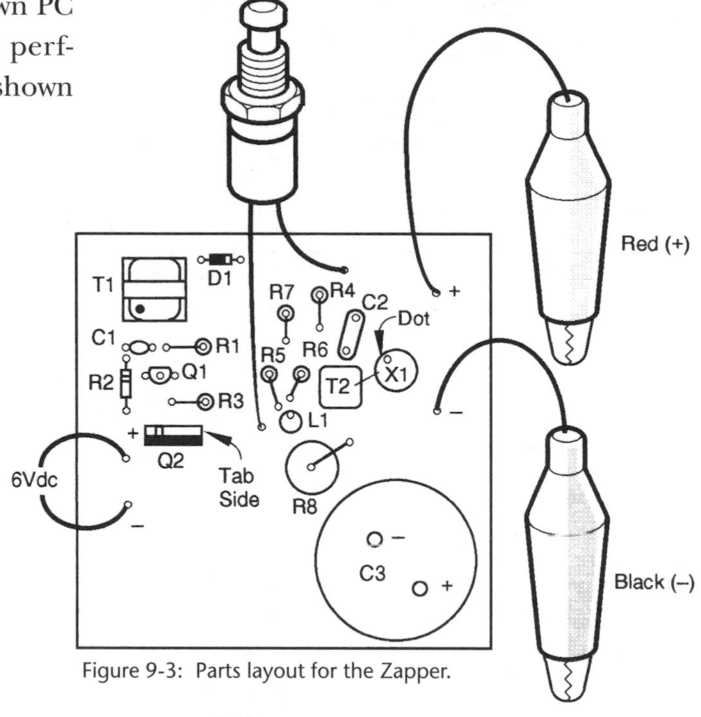

Figure 9-3: Parts layout for the Zapper.

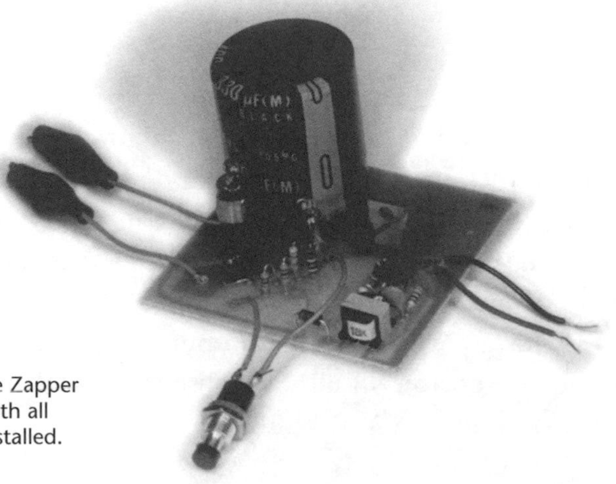

Figure 9-4: The Zapper circuit board with all components installed.

Testing the Zapper

After double-checking your assembly work (especially for correct parts placement), connect 6 VDC (your choice of a lantern battery or four C or D cells in series) to the appropriate points on the board. **KEEP IN MIND THAT THIS CIRCUIT CAN GENERATE OVER 180 VDC!** *Be careful not to come in contact with the board, the output clips, or the terminals of the large electrolytic capacitor!*

After an initial period of up to five minutes (required to initially "form" capacitor C3), the neon lamp should light up orange. If this does not occur, you have probably made an assembly error. In this case, short the two terminals of C3 with the blade of an insulated handle screwdriver and consult the "Troubleshooting" section of this chapter.

To monitor the charging voltage across C3, connect the plus probe of a voltmeter (set to at least 200 VDC) to the cathode of D1 (the banded end) and the negative meter probe to the terminal of switch S1 that goes to resistor R7. You should see the voltage rise rapidly at first and then slowly as it approaches about 160 volts, at which point the neon lamp should turn on. The voltage will continue to rise slowly to about 180 VDC.

If the neon lamp is glowing as it should, you may disconnect the 6 VDC battery and then press the pushbutton switch. If you have assembled the NiCad Battery Zapper properly, X1 will flash white and you may hear a loud "pop." The longer you wait after the neon lamp lights up (up to two minutes), the higher the zapping voltage.

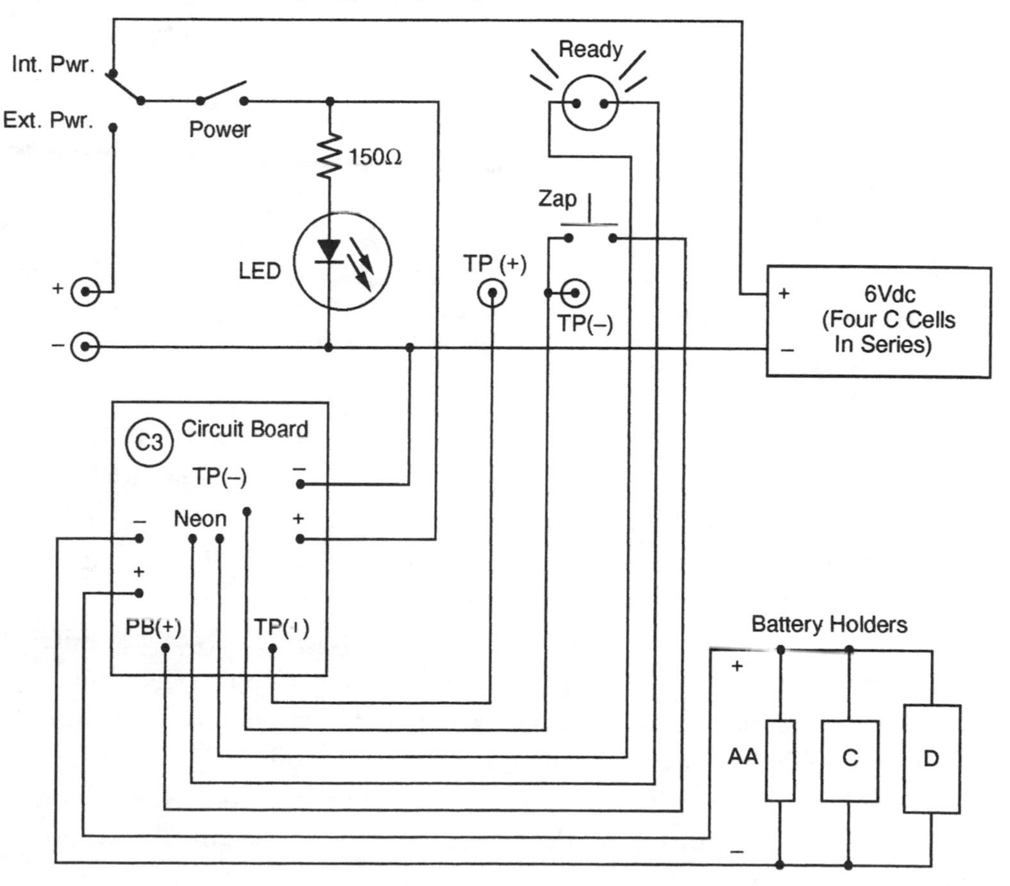

Figure 9-5: Connections between the circuit board and other components.

Packaging

Figure 9-5 shows the wiring connections between the circuit board for the Zapper and the rest of the unit's components. However, I don't like to connect clip leads to a project with high voltage on the loose! I built my Zapper into a large enough cabinet to hold the circuit board and four C-size batteries in an internal holder, as shown in Figure 9-6.

On the cover, I mounted the "external"

parts: the battery holders, the "READY" neon lamp, the "ZAP" pushbutton, and a "POWER ON/OFF" switch. I also added two terminals for "EXTERNAL POWER," an "EXT.PWR/INT.PWR" toggle switch, and two "VOLTAGE TEST" monitoring points.

The wiring and additions are shown in Figure 9-5. The connections between the internal parts and the underside of the cover were made with colored "ribbon cable" to keep the interconnecting wires from getting mixed up.

Figure 9-7 shows the front of my unit, with the placement of the battery holders and labeling of the controls. The circuit board and internal batteries are inside the housing.

Figure 9-6: A look inside the Zapper housing.

Using the Zapper

If you package the Zapper as I did , then you have your choice of internal or external power. If you use internal power, as I did, then you should use either C or D cells. If you use external power, your best choices are a 6-volt lantern battery or a well regulated adjustable DC power supply set for about 6 volts (be sure to observe correct polarity). The Zapper will draw about 100 milliamperes at first, increasing to about 300 milliamps during each charge.

Figure 9-7: Front view of the completed Zapper

Place the cell to be zapped into one of the battery holders. If the battery won't fit into a holder, use clip leads to connect it to the battery holder terminals. Make sure the positive lead connects to + and the negative lead to – ! **If you have a multi-cell NiCad battery pack, you must zap only a single cell at a time.** For details on how to locate a single cell in a multi-cell pack, refer to the "Save That Battery!" section earlier in this chapter.

Once you have the cell connected, set the EXT.PWR/INT.PWR switch to the proper setting, and then put the POWER switch in the ON position. The red LED should light immediately. In about two minutes (up to five minutes if the Zapper hasn't been used for a while), the READY neon bulb should light. Wait about ten more seconds, and then press the ZAP switch (you may have to do this more than once). You should see the xenon tube flash and hear a "pop."

If you do not hear a loud pop, you may not have made a good, firm connection between the clip leads and the NiCad, or the NiCad was not shorted. The Zapper DOES NOT produce a loud pop when a good cell has been zapped, but it does not harm the good cell. Incidentally, the Zapper cannot repair cells in which the electrolyte has been totally released, such as when a cell has been badly overcharged.

The battery or batteries that supply power to the unit should be replaced when it takes longer than five minutes for neon lamp L1 to light.

If you'd like to monitor the internal charging voltage, use a voltmeter set for at least 200 volts connected to the VOLTAGE TEST points, observing polarity. **Be careful not to touch these test terminals with your fingers, since up to 180 volts can be present!**

Troubleshooting

Re-check all resistors for the proper values installed in the proper places. Look for solder bridges across parts (or printed circuit traces). Check for cold solder joints and re-heat, adding solder to any suspect connections.

Referring to Figure 9-3, make sure that both transistors have their flat sides facing the direction shown. Check to make sure that the proper polarities were observed when mounting electrolytic capacitor C3 and diode D1. Check to make sure that xenon quench tube X1 is connected with its white dot closest to capacitor C2, and the white dot lead of transformer T1 is closest to capacitor C1.

If the C3 voltage (measured at the test points) does not reach high enough to light the neon bulb, change resistor R6 to 470 kilohms; this will raise the voltage at the junction of R5 and R6.

Happy Zapping!

At the time this book was written, the parts marked with an asterisk in the parts list were available from Electronic Goldmine, P. O. Box 5408, Scottsdale, AZ, 85261; their toll-free order line was 800-445-0697. They also offered a kit of all parts, except for the enclosure, for $18.75 plus $5 shipping. Pricing and availability may have changed since publication of this book.

PARTS LIST

C1:	0.1 µF disc ceramic capacitor
C2:	0.1 µF Mylar capacitor
* C3:	330 µF 250 V electrolytic capacitor (Electronic Goldmine #G6747)
D1:	1N4007 1 amp 1000V silicon diode
L1:	Neon lamp
Q1:	2N3904 NPN general purpose silicon transistor
Q2:	TIP30 PNP 30 watt silicon power transistor
R1:	6.5 kilohm 1/4 W carbon resistor
R2:	68 Ω 1/4 W carbon resistor
R3:	43 kilohm 1/4 W carbon resistor
R4, R7:	1 megohm 1/4 W carbon resistor
R5:	470 kilohm 1/4 W carbon resistor
R6:	560 kilohm 1/4 W carbon resistor
R8:	1.2 kilohm 1 W carbon resistor
S1:	Normally-open pushbutton switch
* T1:	White-dot inverter transformer (Electronic Goldmine #N1703)
* T2:	Red trigger transformer (Electronic Goldmine #N1700)
* X1:	Xenon quench transformer (Electronic Goldmine #A1048)
Battery holders:	1 each size AA, C, D
Binding posts:	1 red, 1 black
Int.Pwr/Ext. Pwr Switch:	Single-pole double-throw (SPDT)
Power On/ Off Switch:	Single-pole single-throw (SPST)
ON indicator:	Red LED
LED resistor:	150 Ω 1/4 W carbon resistor
Test points:	2 screws and nuts
Miscellaneous:	Circuit board (PC or perfboard), ribbon cable
Enclosure:	Plastic cabinet approximately 7 7/16" x 4 1/4" x 2 3/8"

CHAPTER 10

A Multi-Purpose Digital Tester

This "multi-purpose digital tester" is a really handy "test-almost-anything" gadget that is useful to electronic designers, troubleshooters, and experimenters. Most of the parts are commonly available, and a printed circuit board is not required (although one is shown). Two different, inexpensive "packaging" designs are described in this chapter.

The "multi-purpose digital tester" is basically a three-speed square wave generator that can be used for testing electronic circuits and components. Used as a "signal injector," it can test digital and analog amplifiers. It can also be used, together with a speaker or earphone and a battery, to test various types of components.

Extensive use assumes some electronic background, especially in digital testing. However, its versatility allows even relative beginners to find it useful in testing analog amplifier and broadcast receivers, as well as individual electronic parts. Before getting into the details of "how it works," a brief description of "how it's used" will make it easier to follow the technical explanation that follows.

Description

The "controls" consist of two two-position slide switches and a pushbutton switch. Power is normally supplied by the circuit under test, or an external 9-volt battery.

With the MANUAL/AUTO slide switch in the MANUAL position, simply depress the pushbutton switch whenever you want an output state change from low to high, or high to low. If you want slow-speed operation, you set the MANUAL/AUTO switch to the AUTO position and the FAST/SLOW switch to SLOW. Then the output changes automatically about once or twice each second, and holds each state for about the same time. By setting the FAST/SLOW slide switch to the FAST position, the output "speed" increases to approximately 460 Hertz (Hz, or cycles per second), still a square wave.

For TTL, DTL or CMOS circuits, you clip the red (positive) and black (ground) leads to the power source and, depending on switch settings, the white output lead provides a high or low, or a "slow" or "fast" square wave "clock." This is particularly useful when experimenting with counting or logic circuits and you want to slow things down so you can see what's happening.

You can trigger all kinds of digital counters and flip-flops, and even linear devices and transistors, with this extremely simple one IC device. For testing amplifiers and radios, the square wave is like a sine wave rich in harmonic frequencies that go up into the high kilohertz (kHz) range, so the signal even goes through broadcast band AM and FM tuned circuits.

Circuit Operation

The schematic of the Digital Tester is shown in Figure 10-1. Two of the six sections of the CMOS 4069 hex inverter IC are used in a bistable circuit; two sections are used in an astable multivibrator circuit; and two sections are used as buffer/drivers for the output.

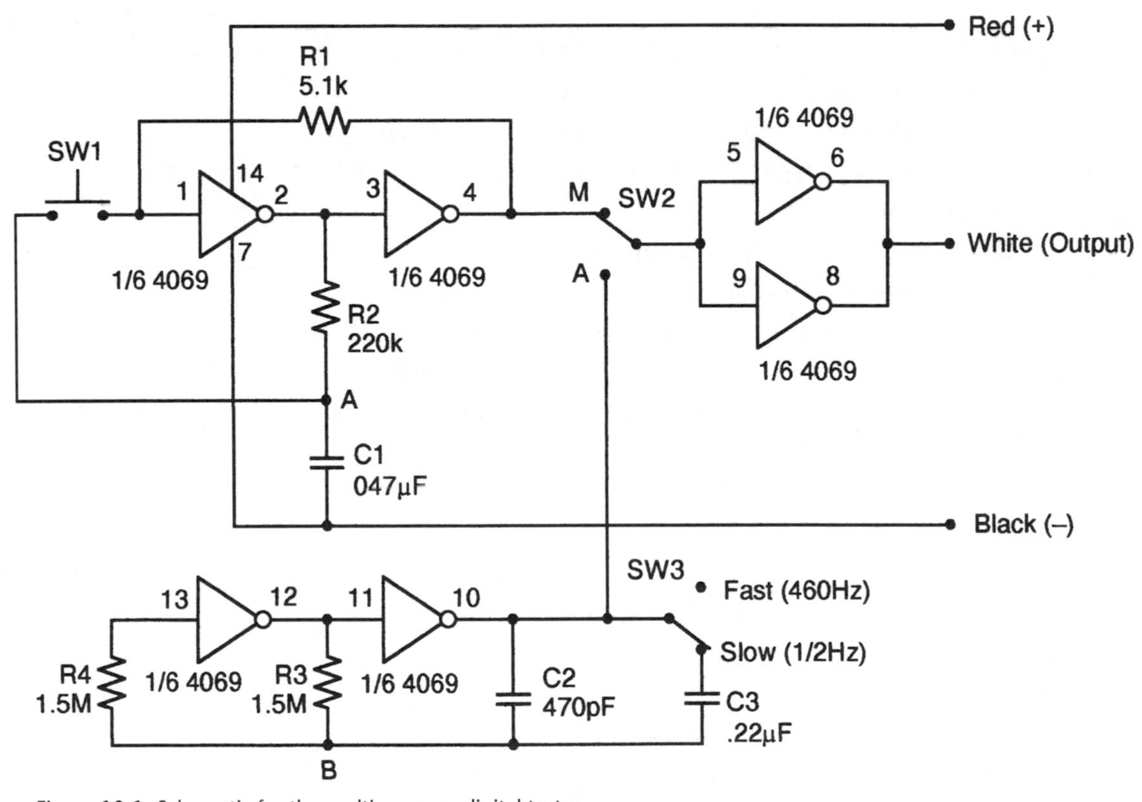

Figure 10-1: Schematic for the multi-purpose digital tester.

IC pins 1, 2, 3, and 4, together with resistors R1 and R2, capacitor C1, and pushbutton switch SW1, form the bistable circuit. Actually, this is a latched flip-flop arrangement which changes state, and holds the "new" state each time the pushbutton is pressed. If the pin 4 output is low, then pressing the button changes pin 4 to high. The next operation of the pushbutton will drop pin 4 to low.

To describe the circuit action, let's assume pin 1 is low and pushbutton switch SW1 is open. Because this is an inverter, pin 2 is high. Since pin 2 is directly connected to pin 3, pin 3 is high, and the inverted output at pin 4 is low. The circuit is stable in this condition, and C1 is charged to almost the supply voltage through R2. Now, when SW1 is closed, the positive charge on C1 is applied to pin 1, which instantly goes high; pins 2 and 3 go low and pin 4 goes high. As long as SW1 is held closed, a voltage divider to the low at pin 2 is formed at Point A, with high voltage fed through R1 from pin 4. Since R1 is such a low value compared to R2, the voltage at Point A and pin 1 is held high.

However, when SW1 is released (opened), Point A is no longer held high, and C1 discharges through R2 to the low at pin 2. The RC time constant keeps C1 high long enough to eliminate the effect of switch contact bounce, so this circuit is fully "debounced." Since pin 1 is still high, there is no change of state when SW1 is opened; pin 4 is still high.

At the next closure of switch SW1, pin 1 sees "ground" at Point A (since C1 is discharged) and pin 1 instantly goes low. Pins 2 and 3 go high, with pin 4 going low. As long as the switch is closed, R1 holds pin 1 low, since it is the lower leg at Point A of the voltage divider formed by R2 and R1 being fed a high voltage from pin 2. When the switch opens, C1 charges through R2, again taking enough time to debounce the switch; pin 1 stays low. We're back where we started!

When switch SW2 is placed in MANUAL (M position), the high or low on pin 4 is applied to pins 5 and 9 of two sections of the IC, which are connected in parallel to increase the current available at the output. These invert and isolate the output so even a low impedance, such as an 8 ohm speaker, won't load down the circuit.

IC pins 13, 12, 11, and 10, together with resistors R3 and R4 and capacitors C2 and C3, form the astable "two-speed" square wave oscillator. To illustrate circuit action, let's assume that, at a particular instant, pin 13 is low. Therefore, pins 12 and 11 are high and pin 10 is low. With selector switch SW3 in the "slow" (S) position, capacitors C2 and C3 are connected in parallel (so their capacitances add together) between pin 10 and Point B.

If pins 12 and 11 have just gone high, capacitors C2 and C3 are charging through R3, thus the voltage at Point B is increasing. When the voltage at this point reaches transfer voltage, pin 13 snaps high, pins 12 and 11 go low and pin 10 goes high. Now the capacitors discharge through R3 until Point B drops below threshold voltage. Pin 13 snaps low back to where we started.

Resistor R4 is connected from the timing circuit (R3, C2, C3) to pin 3 to limit current to the built-in CMOS diode protection gate input circuit. The action keeps repeating at a rate determined by the setting of switch SW3-SLOW or FAST. When in the FAST position, the larger capacitor C3 is removed from the circuit; since the total capacitance is now much lower, and the charge/discharge times are shorter, the circuit changes state faster.

When MANUAL/AUTO switch SW2 is placed in the AUTOMATIC (A) position, the bistable circuit output is disconnected and the output of the multivibrator (FAST or SLOW, as set by SW3) is connected to the buffer/inverters, which trigger the circuit output (white) lead.

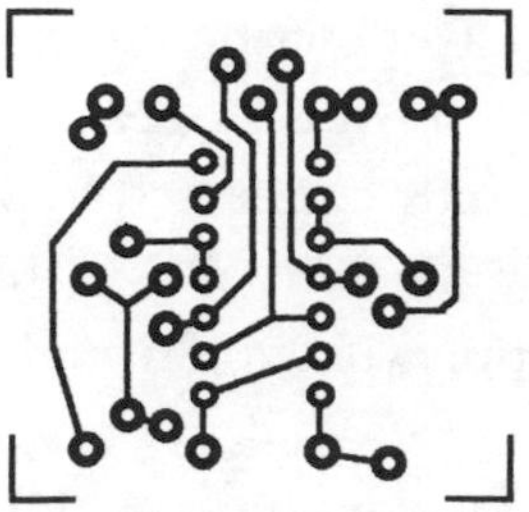

Figure 10-2: Printed circuit board layout for the tester.

Construction

This circuit can be assembled on perforated board or you can make a printed circuit board as shown in Figure 10-2. There is nothing critical about parts layout if you use perforated board; if you use the printed circuit, Figure 10-3 shows the parts layout.

It's a good idea to use a 14-pin integrated circuit socket instead of soldering the 4069 IC directly to the board. Solder the socket and component leads carefully, using 0.031 diameter resin-core solder and a fine-tipped 25–50 watt soldering iron. Clip off the excess leads and examine the soldering carefully for unintentional solder "bridges." When you plug the 4069 IC into its socket, be sure the notch is oriented as shown.

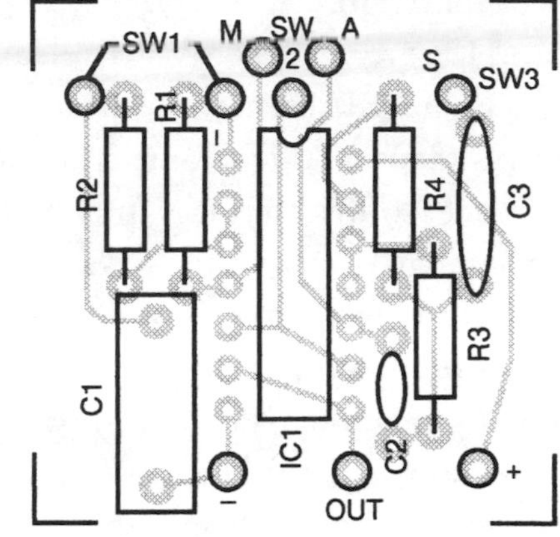

Figure 10-3: Parts placement and layout on the circuit board.

Packaging

There are several ways to house this circuit. Any small plastic box will do. I made two versions, one in a 35mm plastic film container, and the other in a Fuji Slide box as I described back in Chapter 6.

The 35mm plastic film container—the kind that holds a roll of 35mm film—is made from a soft plastic easily cut with a razor blade or X-acto knife, and the cap snaps firmly on the formed rim of the can. To prepare a film container as the case for this circuit, cut a hole in the bottom large enough for the three-conductor cable, and cut holes in the cap for the three switches. The pushbutton switch is held in position with the large nut that comes with the switch. If you use a film container as the case, you'll need subminiature slide switches (see parts list) that are held to the cap with #2-56 screws and nuts. Figure 10-4 shows the layout of the switches. If you use a slide box, you can put all the switches on the lid and the board with all the parts inside. The wiring is less crowded using the slide box,

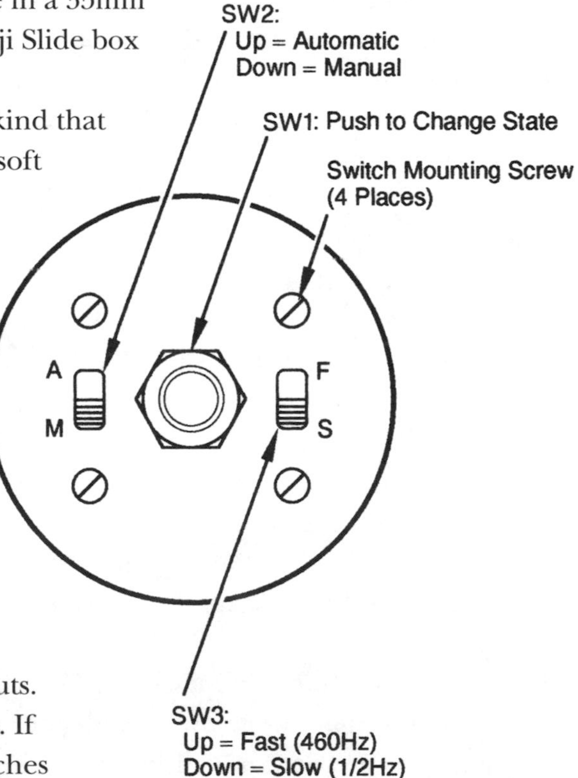

Figure 10-4: Switch location placement atop a 35mm film container housing.

and you can use standard size slide switches.

In either case, next wire the board assembly to the switches as shown in Figure 10-5 (if you use the PC board the legends are shown in Figure 10-3), or based on the Figure 10-1 schematic. Be sure that none of your wires and switch terminals are shorted together.

Now connect one end of the three-wire cable to your circuit board, solder the clips on the

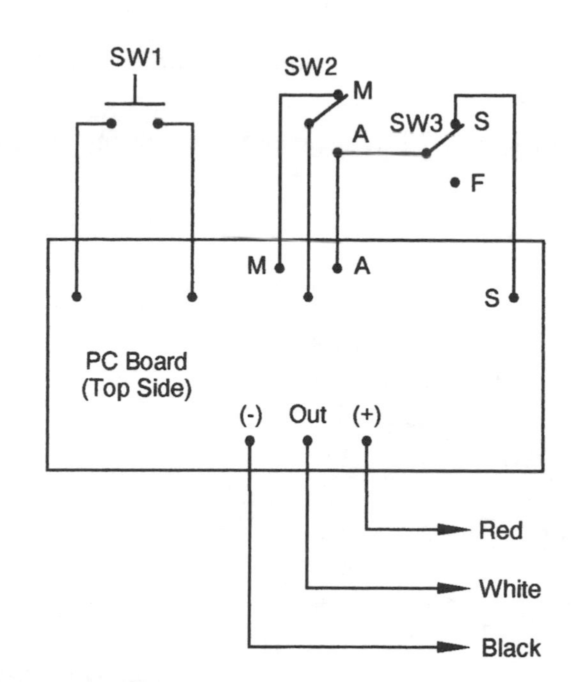

Figure 10-5: Wiring connections between the circuit board and switches.

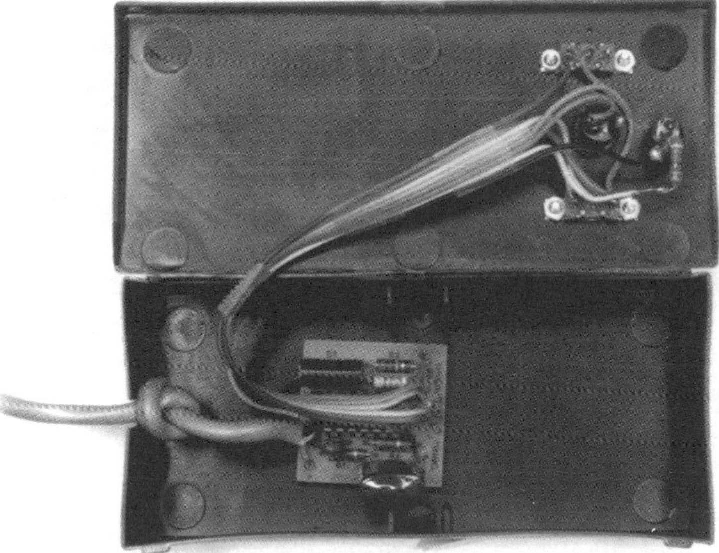

Figure 10-6: Mounting of the circuit board and other components inside a slide box.

other ends of the wires, slip the proper colored insulator over each clip, and your multi-purpose digital tester is complete! Figure 10-6 shows the interior connections of a unit installed inside a Fuji Slide Box, and Figure 10-7 shows the exterior appearance of a final version that I assembled inside a Fuji Slide Box.

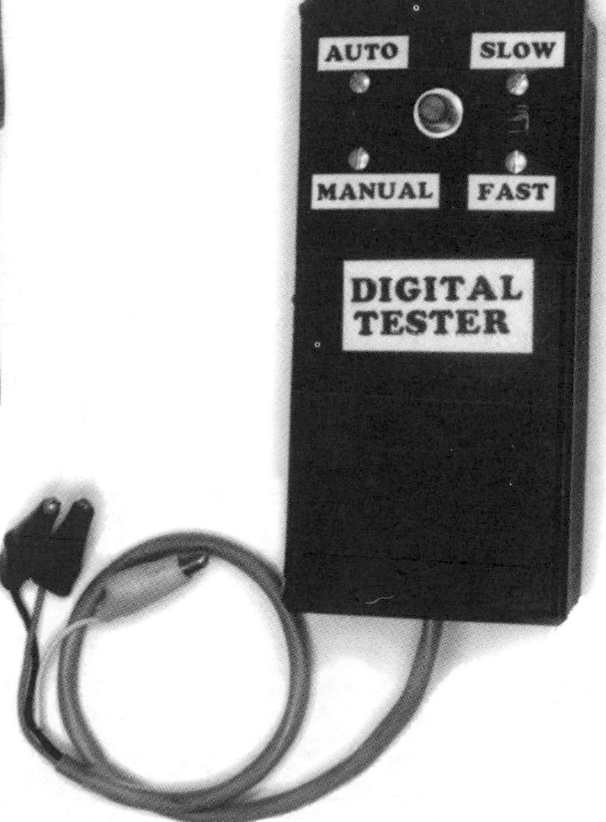

Figure 10-7: Exterior view of a unit built inside a slide box.

Figure 10-8 shows a perfboard version of this circuit installed inside a 35mm plastic film container. The wiring connections are made to the switches positioned as seen in Figure 10-4. One advantage of using a 35mm film container is that the completed unit fits nicely into the palm of your hand, with all controls within easy reach of your thumb. Figure 10-9 illustrates this.

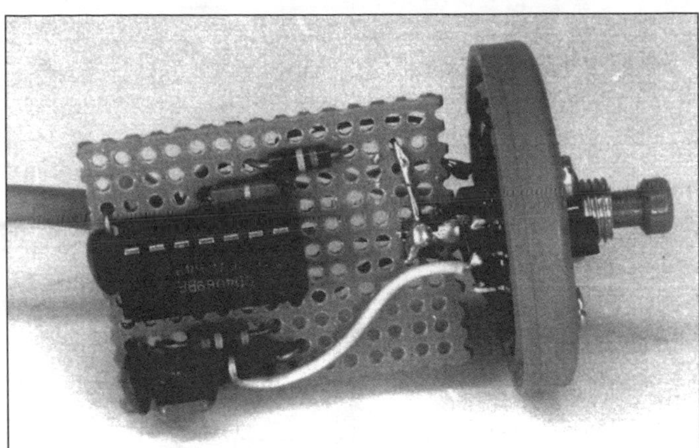
Figure 10-8: Circuit board to fit inside a 35mm film container.

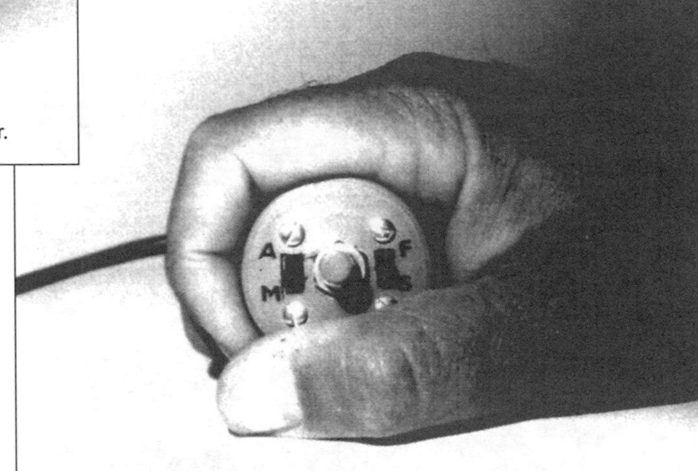

Figure 10-9: A 35mm film container housing fits easily into the palm of your hand.

Using the Digital Tester

The basic use for this digital tester is to trigger digital circuits; in other words, a digital pulser. Connect the red clip to circuit positive voltage, the black clip to ground, and use the white clip as the "trigger," setting the switches for the operating mode you desire. Figure 10-10 shows the bistable and astable signal level changes.

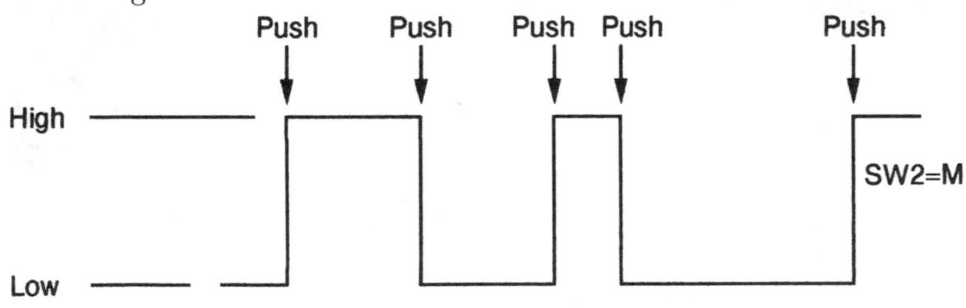

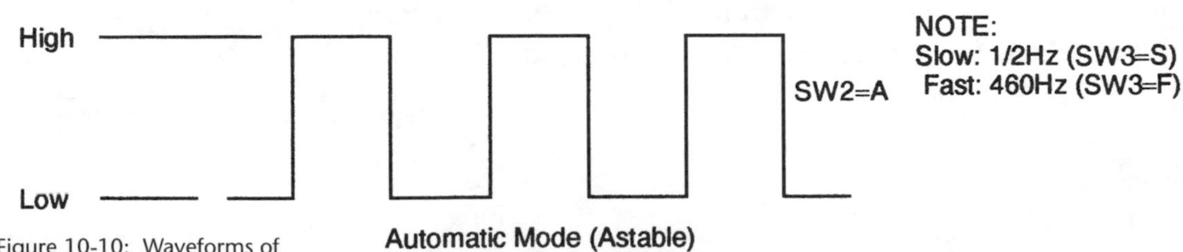

Figure 10-10: Waveforms of manual and automatic modes.

In the MANUAL mode, you can trigger a counting circuit, for example, and hold the desired state as long as you want. You can determine if the circuit operates on a positive-going or negative-going pulse edge with a voltmeter, scope, or LED status indicator connected between the white and black leads.

Figure 10-11 shows you how to monitor the output signal with an optional LED (light emitting diode) and a resistor. In the MANUAL mode, the LED lights when the output is high, it blinks at the cycle rate when in SLOW, and appears to be on all the time when in FAST mode (although it is only on half the time).

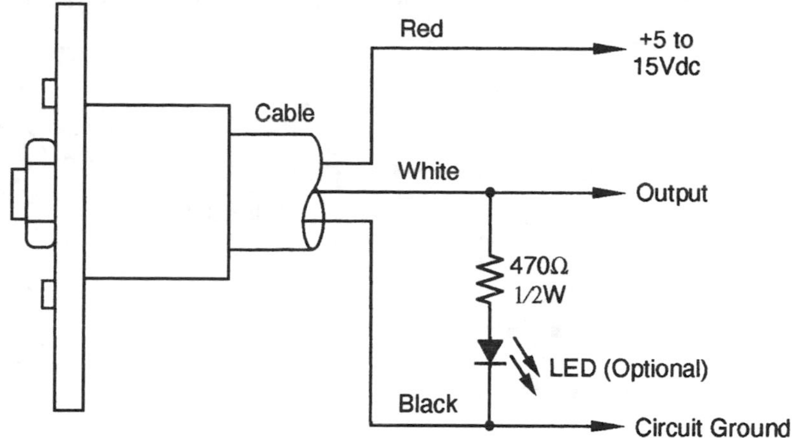

Figure 10 11: Adding an LED to indicate output state.

Actually, if you want to add the LED and resistor to your digital tester as a status indicator, you can mount the LED in the top of the case permanently; however, the current requirement for the circuit will go from about 1 milliampere (mA) without the LED to over 5 mA with the LED.

The circuit shown in Figure 10-11 is also useful for troubleshooting if your digital tester doesn't seem to be working properly, Connect the positive terminal of a typical 9-volt battery to the red clip lead, and the negative terminal of the battery to the black clip lead. An LED connected as shown in Figure 10-11 should light on-and-off depending on the position of switches SW1, SW2 and SW3, as described earlier. If SW2 is in the Manual mode, pressing SW1 should cause the LED to blink on and then off with the next press, changing state with each press of SW1. If this does not happen, R1, R2 and C1 are suspect. If SW2 is in the Automatic mode and the LED does not blink slowly in the SLOW mode, or seem to be on all the time in the FAST mode, then R1, R3, C2, and C3 are suspect. If the LED does not operate at all, then the integrated circuit is suspect, especially the connections at pins 5, 6, 8, and 9.

Your digital tester can also be useful in testing amplifiers, radios and many electronic components. In typical transistor radios, which operate on 6, 9, or 12 volts, the digital tester can be powered by the radio power supply. If it's inconvenient to do this, use a standard 9-volt transistor radio battery to power the digital tester, being sure to connect the negative lead of the battery to the circuit ground of the radio you are testing. In series with the output lead of the tester, use a 0.01 mF capacitor of sufficient voltage to isolate the circuits, and "probe" the radio circuits using the FAST setting of the digital tester as a signal injector. It has sufficient power to drive the speaker, so start there and move backwards through the circuitry until you find a "dead" stage. At that point, voltage and continuity checks can be used to isolate the bad part.

Using a 9 V battery and a small 8 Ω speaker or earphone, you can put the digital tester to work as a component tester when operating in the FAST mode. See Figure 10-12. When testing resistive devices, the sound from the speaker will be loudest at lowest resistance, and you will still be able to hear some sound from the speaker with up to 15 kilohms in series with the speaker.

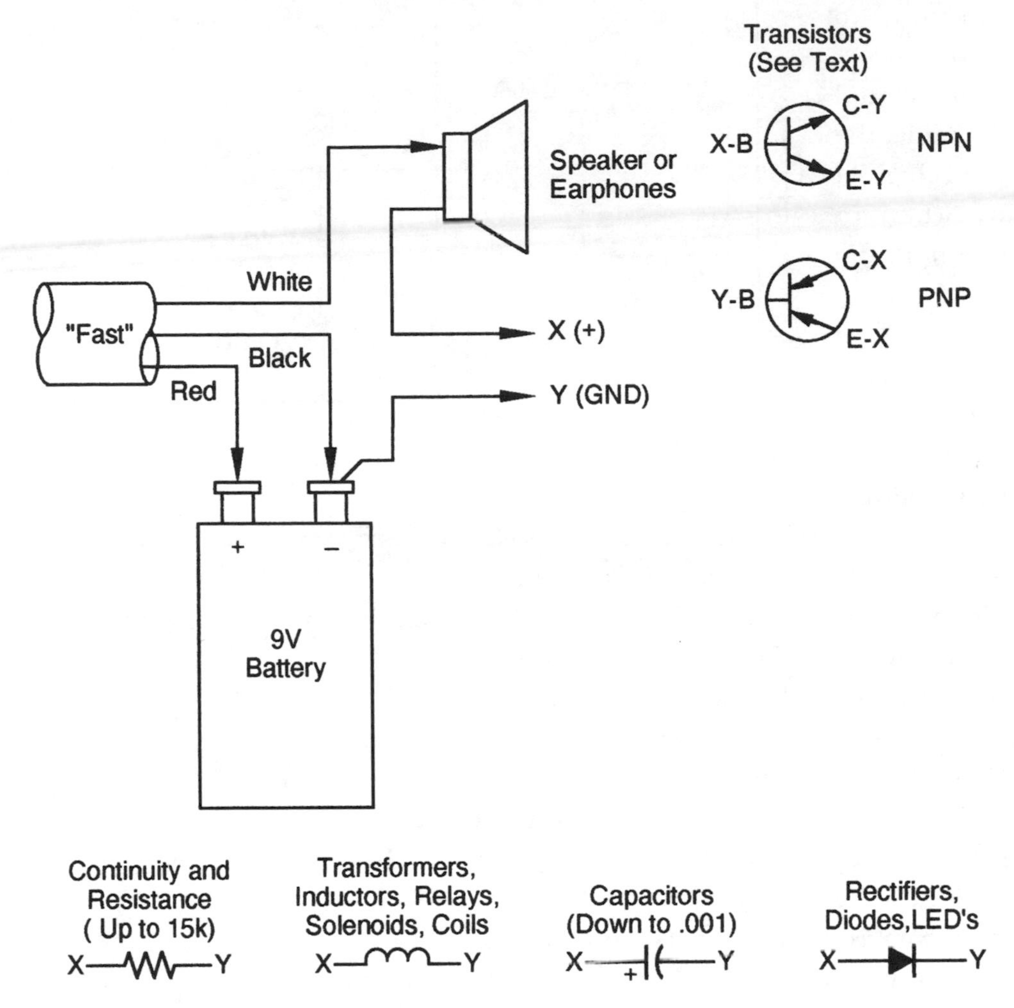

Figure 10-12: A battery and speaker can help you test various kinds of components.

When testing capacitors, the larger the capacitance, the louder the sound through the speaker. Capacitors as small as 0.001 microfarads (1000 picofarads) will still be audible. When testing polarized capacitors (electrolytic, tantalum, etc.) be sure test point "X" is connected to the positive lead.

The rectifier/diode/LED test also determines polarity, since the speaker is silent if the connection is reversed from that shown. An LED under test will light when connected as shown, with current limiting provided by the IC output and the speaker impedance.

Transistors are tested as if they were composed of two diodes with a common base. First determine which leads are the base, the emitter and the collector. Clip the "X" test lead to the base and alternately connect "Y" to the collector and emitter. Speaker sound in both cases tells you the transistor is a functioning NPN type. If the speaker is silent, reverse the leads so test lead "Y" is connected to the base, and "X" alternately connected to the collector and emitter. Sound now means you have a functioning PNP transistor.

Many other components, such as switches, incandescent bulbs, photocells, earphones, some microphones, potentiometers, patch cords, etc., can be tested for continuity. Also, you can use any battery from 5 to 15 volts for these tests. If you choose, you can build all the circuitry in a larger case to include the digital tester circuit, battery, LED and speaker, and have a portable "universal" tester! Simple, small, inexpensive, portable, easy to build and very versatile in its applications, this multi-purpose digital tester can become one of your most useful items of test equipment.

NOTE: *Mouser Electronics part numbers are shown in parentheses in the parts list for the relatively uncommon parts. Mouser's address is Mouser Electronics, 2401 Hwy. 287 N., Mansfield, TX 76063-4827; telephone 800-346-6873. Availability of these items through Mouser may have changed since this book was published.*

PARTS LIST

C1:	0.047 µF 16 WVDC ceramic disc
C2:	470 pF 16 WVDC ceramic disc
C3:	0.22 µF 16 WVDC ceramic disc
IC:	4069 hex inverter DIP integrated circuit
R1:	5.1 kilohm 1/4 W 5% carbon resistor
R2:	220 kilohm 1/4 W 5% carbon resistor
R3:	1.5 megohm 1/4 W 5% carbon resistor
R4:	1.5 megohm 1/4 W 5% carbon resistor
R5:	470 Ω 1/4 W 5% carbon resistor (optional; see text)
LED:	Red diffused light emitting diode (optional; see text)
SW1:	N.O. pushbutton switch, red top
SW2, SW3:	Subminiature SPST slide switch (10SM007); see text
Miscellaneous:	14-pin DIP socket, two 2-56 x 1/4" long screws, round head (5721-256-1/4); two 2-56 hex nut (5721-256); three miniature alligator clips (13AC120); alligator clip insulating boots, black (13AC122); red (13AC121); yellow (white no longer available; 13AC124); 12-inch length of 3-conductor cable (515-6352-12); perforated phenolic board, film container or Fuji Slide Box.

CHAPTER 11

3½ Digit Module Applications

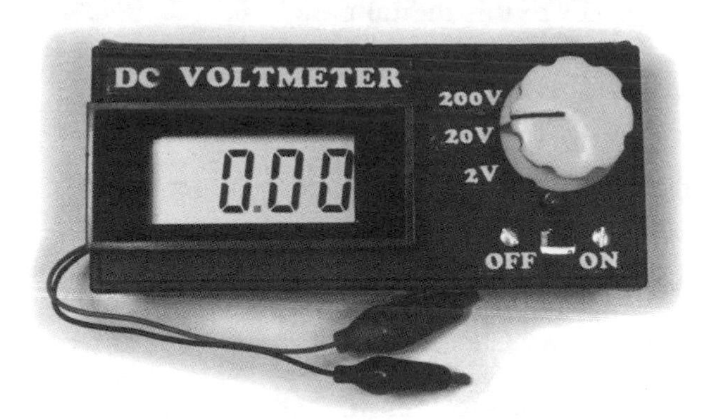

Accurate 3½ digit displays are widely available and inexpensive, letting you substitute a digital readout instead of an old-fashioned analog meter in your projects. In this chapter, we'll see how to construct a small pocket-size DC voltmeter or milliammeter with a 3½ digit display.

The "heart" of the projects in this chapter is a digital voltmeter module with a 40-pin integrated circuit, a 3½ digit liquid crystal display (LCD), and a small number of discrete components. Its part number is PM-128, and it is available from the sources at the end of this chapter. Although essentially a 200 millivolt digital voltmeter, this module, with the proper signal conditioning, can be used to read various ranges of voltage, current or resistance. It is relatively simple to have it read a specific single range, and switching can be added to perform multiple ranges and even multiple functions. Most common low-cost digital multimeters (DMMs) use the same integrated circuit, a 7106, as the module described here.

Analog or Digital?

Sometimes the use of an analog meter makes more sense than a digital display. If the value being measured or monitored changes slowly, or you are looking for a maximum or minimum (null) reading, an analog meter, which reads continuously, is easier to interpret than a choppy, constantly changing digital display (which may only capture readings two or three times a second).

Also, if the measurement is changing rapidly, with large excursions, a digital display has considerable lag and can vary from confusing to useless. In this case an analog meter with fast response is preferable.

However, for most common uses, a properly calibrated digital display provides greater precision and accuracy, and is easier for inexperienced users to read and monitor. The module described here is particularly appropriate for use in a device dedicated to a particular application, or to monitor a specific sensor.

Description

I obtained three of the modules from the three sources listed at the end of this article. Each appeared identical, and came in a small box identifying the unit as "Model PM-128." They each had a data sheet labeled "3-1/2D LCD Digital Panel Meter PM-128." And they were all made in China. Although some sources use their own catalog number, I will refer to this digital panel meter in this chapter as the PM-128.

The PM-128 assembly consists of two pieces: a circuit module that contains the 7106 integrated circuit, the display, and all the necessary on-board discrete components; and a removable plastic front frame.

The 2.6-inch wide by 1.65-inch high x 0.5-inch thick module includes a 1.75-inch wide x 0.75-inch high liquid crystal display with 0.5-inch high digits. The 2.75-inch wide by 1.25-inch high x 0.3-inch thick front frame has two long studs that extend through holes in the module.

The idea is to place the frame in front of a panel (up to 0.25 inches thick) with a 1.75 x 0.75-inch viewing slot and two 0.125-inch diameter holes for the frame studs, them mount the module behind the panel on the studs, and lock the module down to the panel with the two supplied nylon nuts. The panel is simply "sandwiched" between the front frame and the module. It's very easy and practical!

Features

The PM-128 is a fully assembled, ready-to-use LCD digital voltmeter that reads 199.9 millivolts (mV) DC full scale with no additional parts, using a 9 to 12 volt DC source. Using only two resistors, you can change the full-scale reading to 1.999 volts DC, 19.99 volts DC, or 199.9 volts DC. The decimal point position is selected by an on-board jumper.

Using precision resistors, the accuracy is +/–0.5%. Since the input impedance is greater than 100 megohms, the use of this meter to read voltage has virtually no effect on the circuit it is measuring (unlike many analog meters that can have drastic effects!). Automatic polarity indication is provided (so you don't have to be sure your test probes are plus-to-plus), with a guaranteed zero reading when there is no input voltage present.

Depending on the external discrete components and sensors used, the PM-128 can be used not only as a voltmeter, but a thermometer, PH meter, dB meter, wattmeter, current meter, inductance meter, capacitance meter, light meter, battery tester, and other industrial and domestic uses. It is just a matter of using sensors and circuitry that allow the module to read voltage or current.

3½ Digits?

You may not be familiar with the meaning of "3½ digits." The 3 digit part is obvious, and can read up to the value 999. The ½ digit, in front of the three digits, is simply the number "1" and the minus sign (actually, segments b, c and g of a typical 7-segment digit). Therefore, the full "all-on" display is –1999. Three decimal point positions, based on the maximum reading, are set by three jumper locations.

Under the Hood

The PM-128 uses "dual-slope integration" for its measuring technique. The measurement cycle is divided into two parts. During the first part of the cycle, a capacitor is charged up by a current which is proportional to the input voltage—the larger the voltage, the greater the charging rate.

The charging process operates for a fixed time period based on 2000 clock cycles of an on-board counter. At the end of this time period, the capacitor will be charged to a voltage that is proportional to the input voltage.

The second part of the measurement cycle is based on how long it takes to discharge this capacitor compared to a fixed precision reference voltage. During the discharge period the counter counts up from zero until a comparator produces an output signal when the capacitor voltage falls to zero. The count is then displayed, directly reading the number of cycles during discharge, up to a maximum of 1999. The counter is then reset to zero for the next charge cycle.

The actual value of the capacitor is not important, since the same capacitor is used for both charge and discharge cycles. The same applies to the clock frequency, so long as it remains constant during the full measurement. The clock frequency determines how often the measurement is taken. The PM-128 has two to three readings per second.

What if the input voltage is higher than the maximum of 1999 discharge counts? To show this over-range condition, the display simply shows a "1" as the first digit, and the three following digits are blanked. An additional feature of the PM-128 is that it includes an auto-zero level correction circuit that clamps the display to "000" when there is no input. Negative polarity is automatically displayed with a "–" sign at the left end of the display if you connect a negative voltage to the positive PM-128 input; the actual reading is not effected.

Using the PM-128

The "documentation" that comes with the PM-128 is limited to a small insert with limited data. While this article will expand considerably on the information provided with the PM-128, it is not intended to cover more than a few of the many possible applications.

The 7106 integrated circuit takes up more than one-third of the space. The inputs are Vin and GND, the battery connects to +9 V and –9 V, RA and RB are voltage-range resistor locations, and a jumper to select the decimal point location is placed at either P1, P2, or P3. A single small ADJUST potentiometer is used to calibrate the module. There are mounting studs halfway down each side that are not shown. (Don't worry about the other components connected to the +9 V and –9 V points; those will be discussed in the next section.)

If you want to use the PM-128 in a piece of test equipment, such as reading the voltage of an adjustable power supply, you must first determine the maximum voltage reading. Then, using Table 11-1, you'll find the value of the resistors to be used at the RA and RB locations, as well as the decimal point jumper location.

For example, referring to Table 11-1, if you want the maximum voltage to be 20 volts DC, use a 100 kilohm resistor at RA, a 10 megohm resistor at RB, and use a jumper at P2. Notice that the module comes set to read 1999 with no decimal point selected, no resistor at RA, and a jumper at RB. If you want to set it to read 200 millivolts maximum, you only need to put a jumper at P3 to place the decimal point for a 199.9 reading. More than likely, however, for a power supply you'll probably want the 2V, 20V or 200V maximums.

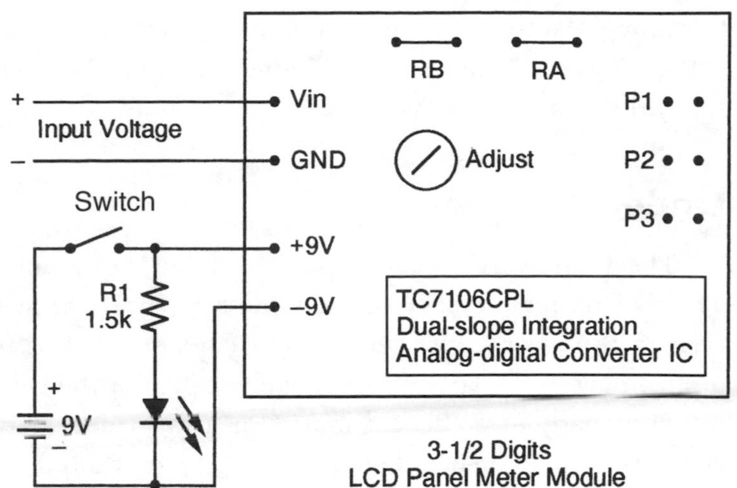

Figure 11-1: PM-128 schematic.

DC Voltage	RA	RB	Decimal Point
200 mV	---	Jump	Jump P3
2 V	1 MΩ	10 MΩ	Jump P1
20 V	100 KΩ	10 MΩ	Jump P2
200 V	10 KΩ	10 MΩ	Jump P3

Table 11-1: Setting various maximum voltage ranges.

Once you've decided on the maximum reading and set the RA and RB values, and set the decimal point jumper, you add 9 to 12 volts DC to power the module. Calibration then involves inputting a known voltage somewhere around mid-scale and adjusting the potentiometer for the correct reading.

A Simple Digital Voltmeter

To illustrate a dedicated use for the PM-128, I built a pocket-sized digital voltmeter. Figure 11-1 shows the circuit (now we'll find out what those other components are used for!), and Figure 11-2 shows the completed unit.

Although the digital display is "on" when battery power is applied, I felt the need for an additional "on" indicator, since it is easy to overlook the display being on when not in use, thus draining the battery. I simply used a red LED and dropping resistor R1 so the LED would light whenever the switch was closed.

Figure 11-2: Configured for a 20VDC maximum reading, the PM-128 module fits nicely in a Fuji 35mm Slide Box, together with a switch, battery, LED, and resistor. The clip leads connect to input voltage.

Granted, this uses a little more battery power, but makes it very obvious when the voltmeter is being powered.

I chose 20 volts DC as my maximum scale, since this reads most common DC voltages such as typical dry cell or alkaline AAA, AA, C, and D batteries, 9-volt radio batteries, 12-volt car batteries, battery packs used in computers, computer power supply voltages, and DC wall-plug converter voltages.

I was able to use a Fuji 35mm slide box (a small plastic box which photo finishers usually provide to hold 36 slides) to house the module, battery, switch, resistor, and LED. This box is made from thin, flexible plastic, making it relatively easy to drill holes for the switch, LED, and the necessary rectangular module viewing area.

The whole assembly is only 5 x 2 x 1.2 inches, and the hinged lid snaps open or closed easily. I chose to have test leads with alligator clips on the end, but you could use probe tips.

To calibrate the unit, use a known-accurate digital voltmeter and a DC power source somewhere around the midrange of the module you'll be calibrating. For example, if your module project is set to read 20 volts DC maximum, you could use a 9-volt battery as a power source.

Monitor the power source voltage with the accurate digital voltmeter and connect the power source to the input of your module project. In effect, you have two voltmeters in parallel. Now adjust the module potentiometer until its display reads the same as the known-accurate voltmeter. The module display should now "track" well at other voltages within the range you set.

Reading Other Voltages

With the 20 VDC digital voltmeter completed, I wanted to see if I could make a multi-range voltmeter (2 volts, 20 volts, 200 volts) in the same size cabinet. This was considerably more of a challenge, since it required a small two-pole three-position switch to select the range and some precision resistors.

Figure 11-3 shows the schematic diagram for this unit. I found the small switch salvaged from an old tape recorder, and was able

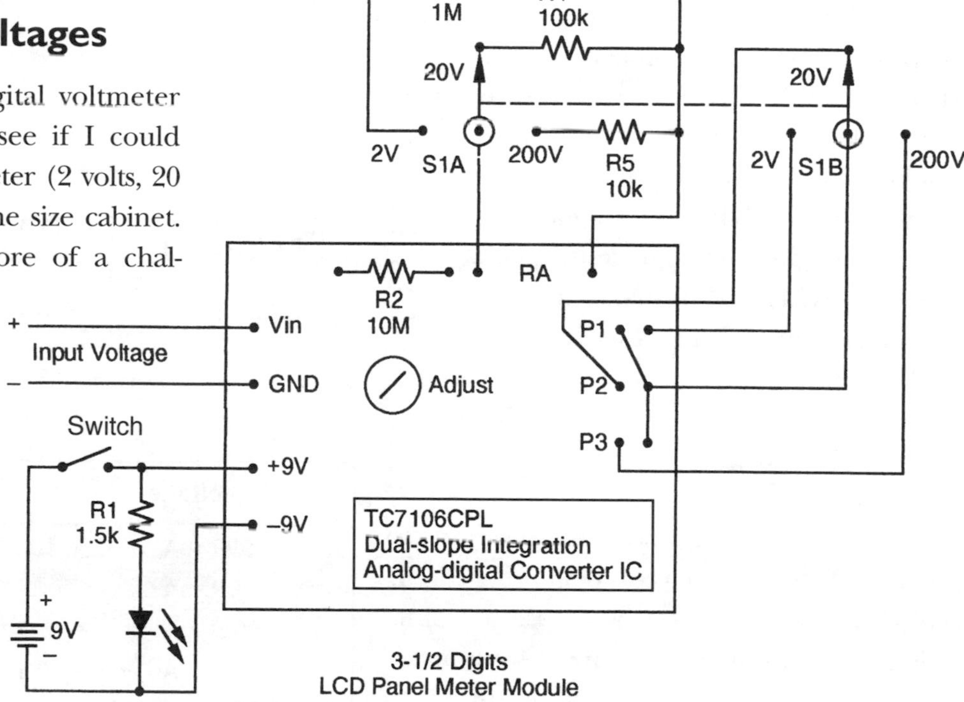

Figure 11-3: Schematic for multi-range voltmeter.

to cram everything, as shown in Figure 11-4, in the same Fuji box. Figure 11-5 shows how the unit looks from the outside.

Resistors R3, R4, and R5 should be precision resistors (1% or better) since you will only be able to calibrate on one range, and the accuracy on the other ranges will depend on the precision of the other resistors. R2, which is used on all ranges, is not as critical.

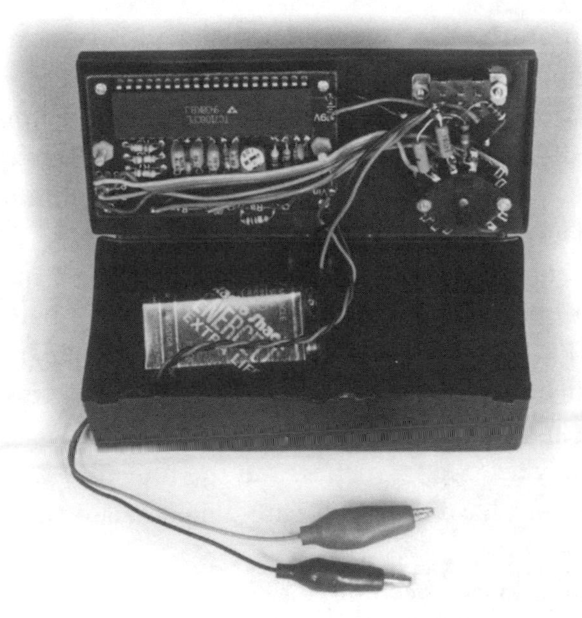

Figure 11-4 (left): The PM-128 module and all other parts but the battery are mounted on the lid of the Fuji Slide Box. This is a tight fit, and requires a fairly small rotary switch.

Note that the common lead in the P1/P2/P3 jumper area, for setting the decimal point, is built into the module, although it is not mentioned anywhere in the module instructions. After construction, connect the battery and calibrate with a known voltage on the mid-range scale, as described previously.

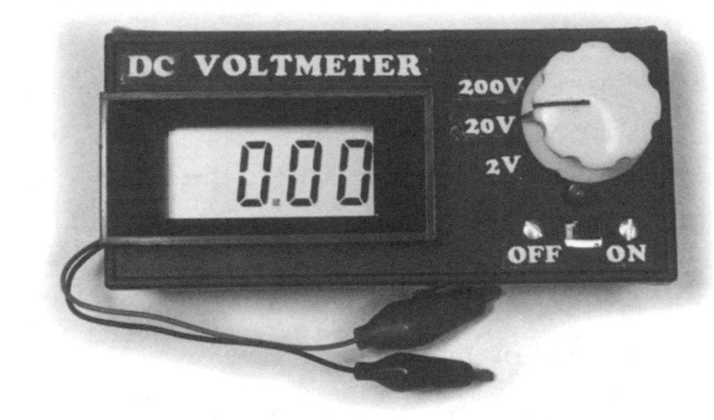

Figure 11-5 (below): Using the PM-128 module with a two-pole three-position switch and some added resistors allows you to make a three-range DC voltmeter.

Reading DC Current

In order to read current with a voltmeter, you must measure the voltage drop across a known resistance. Figure 11-6 and Table 11-2 provide the information you'll need. Figure 11-7 shows that you will be placing the module input probes in series with the circuit being measured, forcing current through resistor Rx.

You decide on a maximum DC current reading and then use Table 11-2 to determine the proper value for Rx and the decimal point jumper location. Table 11-2 assumes the PM-128 module is set for a 200 mV reading.

Be aware that the lower the value of Rx, the less effect on the circuit being measured. Probably the best choice is to use a 1-ohm resistor for Rx (and a P3 jumper) for a maximum reading of 199.9 mA, easily sufficient

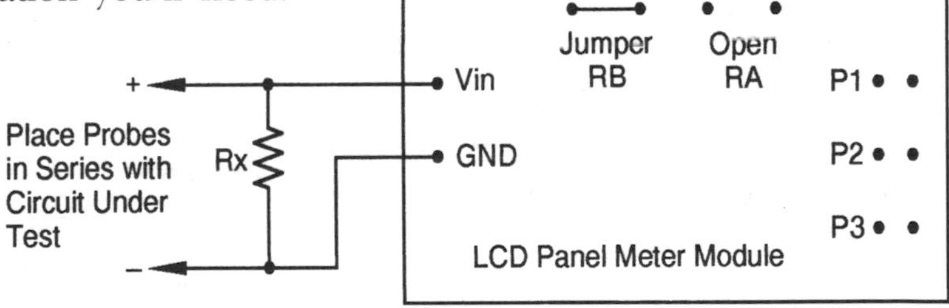

Figure 11-6: Reading current with the voltmeter.

DC Voltage	Rx	Decimal Point
1.999 mA	100 Ω	Jump P1
19.99 mA	10 Ω	Jump P2
199.9 mA	1 Ω	Jump P3
1.999 A	0.1 Ω	Jump P1

Table 11-2: Setting various maximum current ranges.

for most electronic circuits. If you attempt to measure very low current with a 100 ohm value for Rx, this could itself (since it is in series with the circuit) upset circuit operation. For measuring many automotive devices, even 1.999 amperes would not be sufficient.

Other Uses

Various transducers and sensor devices output current or voltage, and these are relatively easy to adapt for use with the PM-128 module.

For those devices that provide output as a variation of resistance, using the PM-128 module is tricky. You would have to provide a constant current source through the unknown resistance, and measure the voltage drop. This can be done with external circuitry, but measuring resistance is best done with an analog or digital ohmmeter designed for that purpose.

PM-128 Sources

Black Feather Electronics
645 Temple Avenue #7
Long Beach, CA 90814
800-526-3717 or 310-434-5641
Catalog number PM-1: 3 1/2 digit panel meter
$9.95 plus $5 shipping per order.
California residents add sales tax

Circuit Specialists, Inc.
P.O. Box 3047
Scottsdale, AZ 85271-3047
800-528-1417 or 602-464-2485
http://www.cir.com
Catalog number: PM-128: 3 1/2 Digit LCD Panel Meter
$9.90 plus $5 shipping per order.
Arizona residents add sales tax

B.G. Micro, Inc.
P.O. Box 280298
Dallas, TX 75228
800-276-2206 or 214-271-5546.
bgmicro@ix.netcom.com
(No catalog number): LCD digital panel meter
$9.95 plus $5 shipping per order.
Texas residents add sales tax

Price and availability from the sources above may have changed since this book was published.

CHAPTER 12

A Photo-Electric Counter

The "Photo-Electric Counter" can be used at home or in business for counting and security applications. It senses the interruption of light falling on a built-in sensor, and advances a 3-digit display one count with each interruption. It can be built from scratch or from a parts kit.

If you want to know how many people or things pass a certain point, a counter is used. Counters are used on production lines, at public events, and at store entrances, among other applications

Traditional electric and mechanical counters are triggered by switches located in places such as on mats or turnstiles. The switches operate solenoids to advance the digits on the linked rotating drums. Because both mechanical and electric counters are essentially mechanical in operation, they are noisy, subject to failure from wear, and limited in their counting speed.

True electronic counters, on the other hand, are completely silent and have no moving parts to wear out. They use lighted digits to display the count, and can operate at high speed. While some electronic counters use switches to trigger their counting, the counter described in this chapter uses light as its count trigger.

Description

This photo-electric counter has a 3-digit display that counts to 999 before resetting to 000. It uses three integrated circuits to detect and count the interruption of light on a photocell, and shows the total count on three red LED 7-segment digital displays. It is completely silent, has no moving parts (except "Hold" and "Reset" switches), and can count at very high speed. I even used it to count the speed of a four-bladed fan by counting the fan blades interrupting a light beam in a given time; try doing THAT with a mechanical or electric counter!

This circuit can be powered by a regular 9-V battery for occasional or short-duration use, or a DC wall-plug transformer for extended use. A sensitivity adjustment allows use in various light conditions, and the count can be reset or stopped at any time.

If you need a count greater than 999, an "overflow" pulse is provided for clocking additional digits. For example, you can easily cascade additional counters for three additional digits each, as I'll describe later.

Circuit Description

Figure 12-1 shows the schematic of the photo-electric counter. The circuit basically consists of three 7-segment LED displays controlled by a counting circuit composed of two integrated circuits (IC2 and IC3) which are triggered by IC1. Figures 12-2, 12-3, and 12-4 show the pinouts of IC1, IC2, and IC3, respectively.

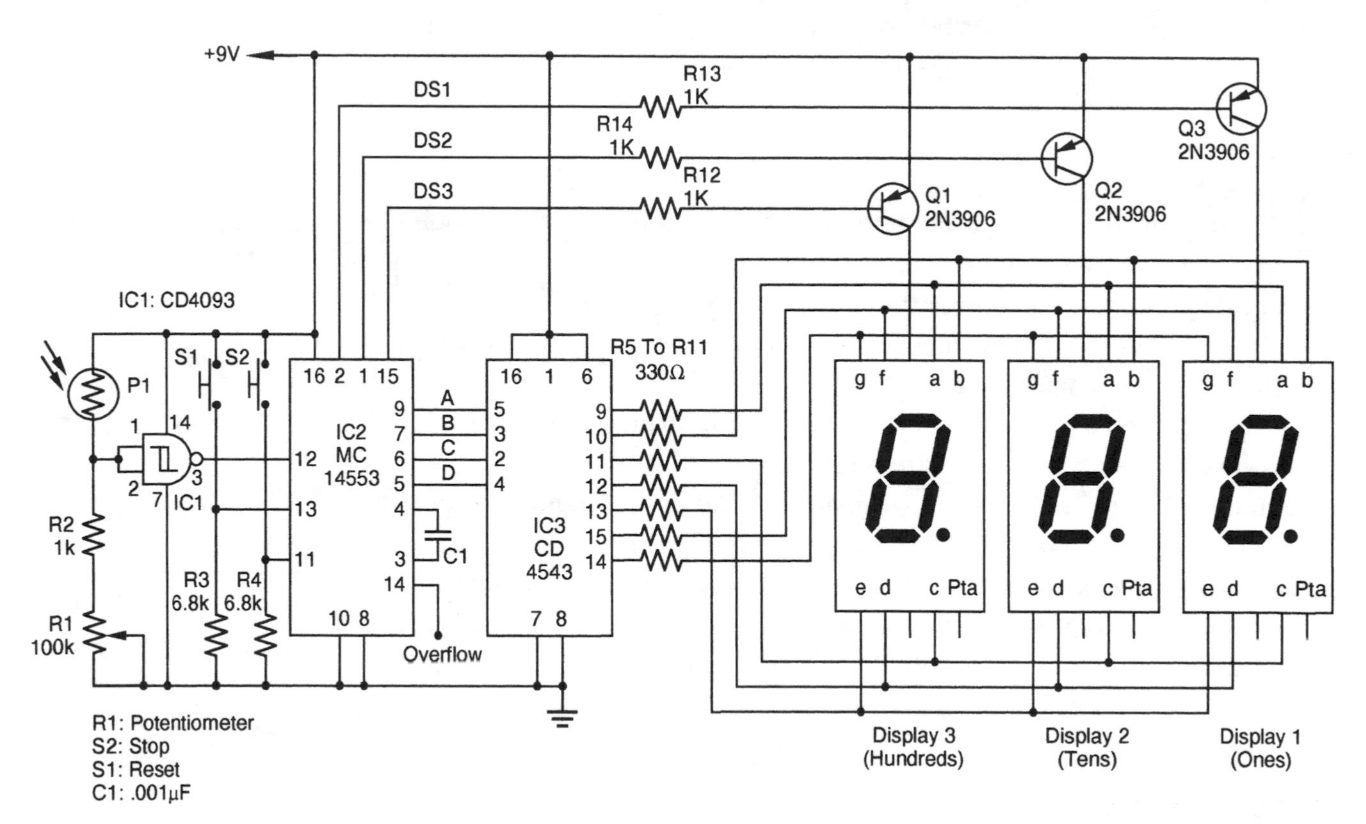

Figure 12-1: Schematic diagram for the photo-electric counter.

Only one section of IC1, a CD4093B quad 2-input Schmitt trigger, is used. This is a NAND circuit, where the output is the opposite of BOTH inputs. That is, if both inputs are low, the output is high, and if both inputs are high, the output is low. This is a Schmitt Trigger device; a change of state is triggered when both inputs increase SOMEWHAT ABOVE one-half the supply voltage, or decrease SOMEWHAT BELOW one-half the supply voltage. The difference between trigger points (almost 2

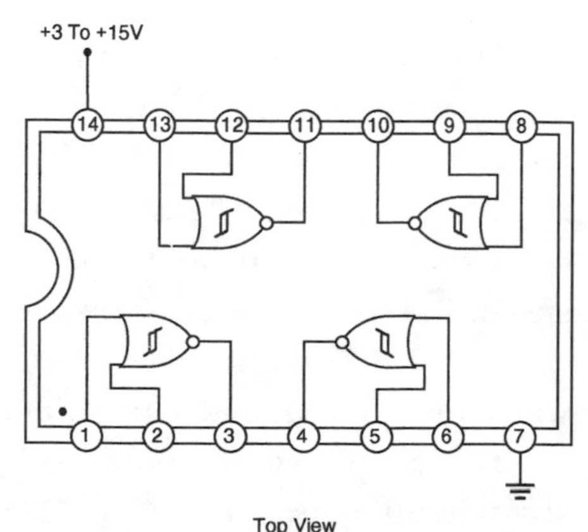

Figure 12-2: Internal diagram of the CD4093B quad 2-input Schmitt trigger IC.

volts with a 9 volt supply) is called "hysteresis," and reduces false response to noisy or slowly changing input voltage levels.

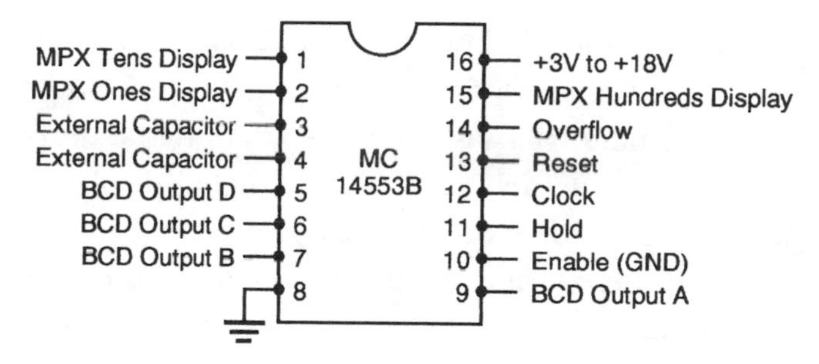

Figure 12-3: Pin functions and connections for the MC14553B BCD counter.

The two inputs, pins 1 and 2, are connected together to a voltage divider at the junction of photocell P1 and resistor R2. Since the photocell has supply voltage at its top, and R2 is connected through potentiometer R1 to circuit ground, the voltage at the IC1 input depends on the resistance of P1 and the setting of R1.

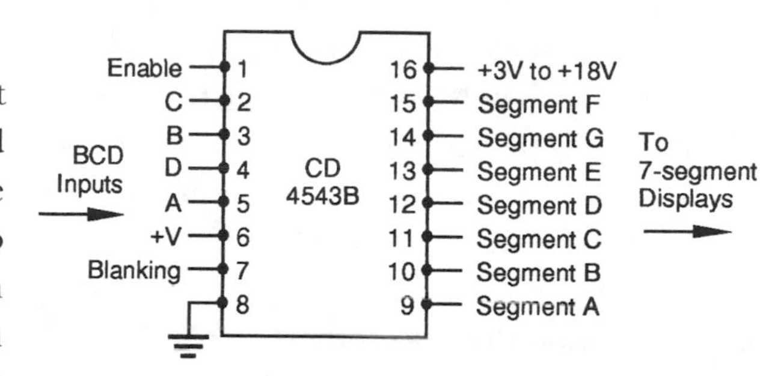

Figure 12-4: Pin functions and connections for the CD4543B BCD to 7-segment latch/decoder/driver.

When little or no light falls on the photocell, it has a very high resistance (several hundred thousand ohms). Since this is considerably higher than the sum of R1+R2, the input to IC1 is at a low voltage, so the output of IC1 is high. However, as light falls on P1, its resistance drops drastically, down to less than 100 ohms in strong light. This causes the voltage at the input to IC1 to increase. When the input to IC1 reaches somewhat above one-half the supply voltage, its output at pin 3 suddenly changes state to low.

Potentiometer R1 lets you set the light sensitivity of the trigger point; the higher the resistance setting of R1, the less the resistance of P1 must drop to trigger IC1, and therefore the greater light sensitivity.

The negative-going voltage at output pin 3 of IC1 acts a CLOCK pulse to pin 12 of IC2, an MC14553B 3-digit BCD counter. This IC is an amazing device. It not only counts upward in BCD ("binary coded decimal")—that is, a four-digit code from 0 to 9 decimal—but it enables the output of this binary code for three digits, one at a time, each for a short duration, using "time multiplexing."

Let's take a look at the concept of "time multiplexing." If neither pushbutton switch S1 or S2 is closed, pins 13 and 11 of IC2 are held low by resistors R3 and R4. This allows IC2 to accept clock pulses. With each incoming pulse, internal counters create the BCD code for each of three digital displays, DS1, DS2, and DS3. As an example, if the total clock count was 319 decimal, then the IC2 internal count for DS1 (units) would be BCD 1001 (9), DS2 (tens) BCD would be 0001 (1), and DS3 (hundreds) BCD would be 0011 (3). (If you don't follow this, don't worry about it, but you may want to brush up on the binary counting system.)

This four-digit binary count for each digit is enabled at pins 9, 7, 6, and 5 (called A, B, C, and D in Figure 12-1), but only for an instant—about 2.5 milliseconds. Why? Because an internal oscillator switches between the three digital BCD codes at a speed dependent upon capacitor C1, which is connected between pins 3 and 4 of IC2.

With a value of about 0.001 microfarads for C1, the IC2 oscillator runs at about 400 cycles per second, changing output state to low at pins 2, 1, and 15 in that order. As you'll see later, this "multiplexing" sets the number to be shown on each of the three digital displays at any instant, although they all appear to be on at the same time.

When the count exceeds decimal 999, an overflow pulse appears at pin 14 (normally low, goes high), which can act as a trigger to another counter, thus allowing higher counts. (More on this later.)

When switch S1 is closed, this puts a high on pin 13, the RESET pin of IC2, and the internal counters are returned to 000. When switch S2 is pressed, this puts a high on pin 11, the HOLD pin of IC2, and counting stops. Unfortunately, since this switch is not "debounced," some jump in the counting may be observed when this pushbutton switch is pressed. (Preventing this would require additional connections to the other NAND circuits of IC1, and a double-throw switch in place of S2.)

The BCD outputs are multiplexed to IC3, a CD4543B BCD to 7-segment latch/decoder/driver, another amazing chip. As each incoming BCD code is input from IC2 to pins 5, 3, 2, and 4 of IC3, internally the BCD code is latched and interpreted to determine which segments of each 7-segment display should be lighted.

Each display has segments designated A to G, as shown in Figure 12-5. These digital displays need to have power to light their segments. They are common anode displays, which means the common to each segment LED is positive voltage, and each segment is looking for a ground (low) to light. The positive voltage is supplied to each display through separate 2N3906 PNP silicon transistors, Q1, Q2, and Q3.

In this circuit, these transistors all have positive voltage applied to their emitters, but they will not conduct current unless there is a lower voltage (negative bias) on their base. Looking at Figure 12-1, you see that pins 2, 1, and 15 of IC2 each go to the base of one of the transistors. When these pins are high, the associated transistor is off.

Display Pinouts

1 - Cathode E
2 - Cathode D
3 - Common Anode
4 - Cathode C
5 - Cathode D.P.
6 - Cathode B
7 - Cathode A
8 - Common Anode
9 - Cathode F
10 - Cathode G

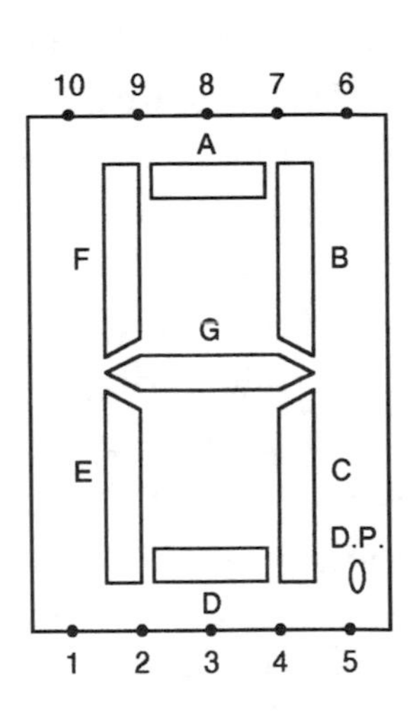

Figure 12-5: Common-anode 7-segment displays.

However, as previously discussed, each of these pins goes low when they are cycled by the IC2 internal oscillator during the time multiplexing process. As each pin goes low, it switches the associated transistor on, allowing it to conduct current to all of the segments of that digit, if the segment sees ground. In other words, when the transistor is on and any segment is low, that segment will light.

Since IC2 is providing the BCD code for each digit simultaneously with the multiplexing of the transistors, and IC3 is latching and converting this code to the 7-digit segment code in step with the multiplexing, each display lights the appropriate segments for the decimal digit to be shown as its

Figure 11-2: Configured for a 20VDC maximum reading, the PM-128

associated transistor is turned on. The sequence is Display 1, Display 2, and Display 3.

Resistors R5 to R11 are used to limit the current to each display segment. Because of multiplexing, each segment is lighted only about one-third of the total time, but the transition between digits is so fast that the eye does not see any flickering. Later, under "Troubleshooting," I'll show you how to slow down the multiplexing frequency for closer observation of this action.

This all may seem confusing on first reading. Read it again, referring primarily to Figure 12-1, and it should make more sense. However, the circuit works even if you DON'T understand it!

Construction

If you purchase the parts kit described at the end of this chapter, all the parts you need—especially the "hard to find" parts—are supplied, and an etched and drilled silk-screened printed circuit board is included. This makes assembly a less-than-one-hour job. Also included with the kit are sockets for the three integrated circuits, a 9-volt battery snap, wire for jumpers, and solder, as well as assembly instructions.

If you desire to build this project from scratch, Figure 12-6 shows the printed circuit board layout, and Figure 12-7 shows the parts layout using this board. While a printed circuit board is not required, it can save a lot of wiring errors, especially in regard to the displays and ICs. The resistors, switches, and transistors are easy to find. However, the integrated circuits (especially IC2 and IC3), the photocell, and the displays may be hard to locate. Possible sources are indicated at the end of this chapter.

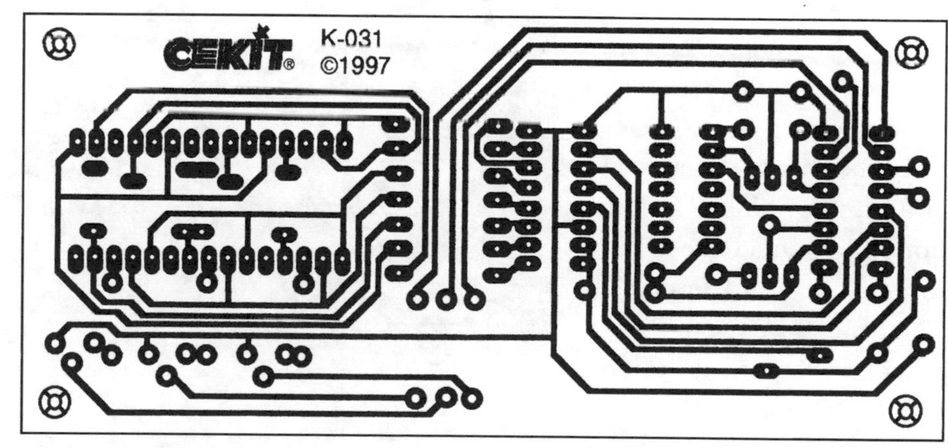

Figure 12-6: Printed circuit board layout for the photo-electric counter.

Use care in assembly regarding parts placement and soldering. Many of the pins are only one-tenth of an inch apart, so be careful to avoid unintentional solder bridges. Be sure to install all eleven jumpers shown silk-screened on the circuit board, including five under the displays! I didn't notice that these jumpers were needed until AFTER the displays were soldered to the board, so I had to figure out where they went and then soldered them on the back (printed circuit side) of the printed circuit board! Figure 12-8 points out these jumpers.

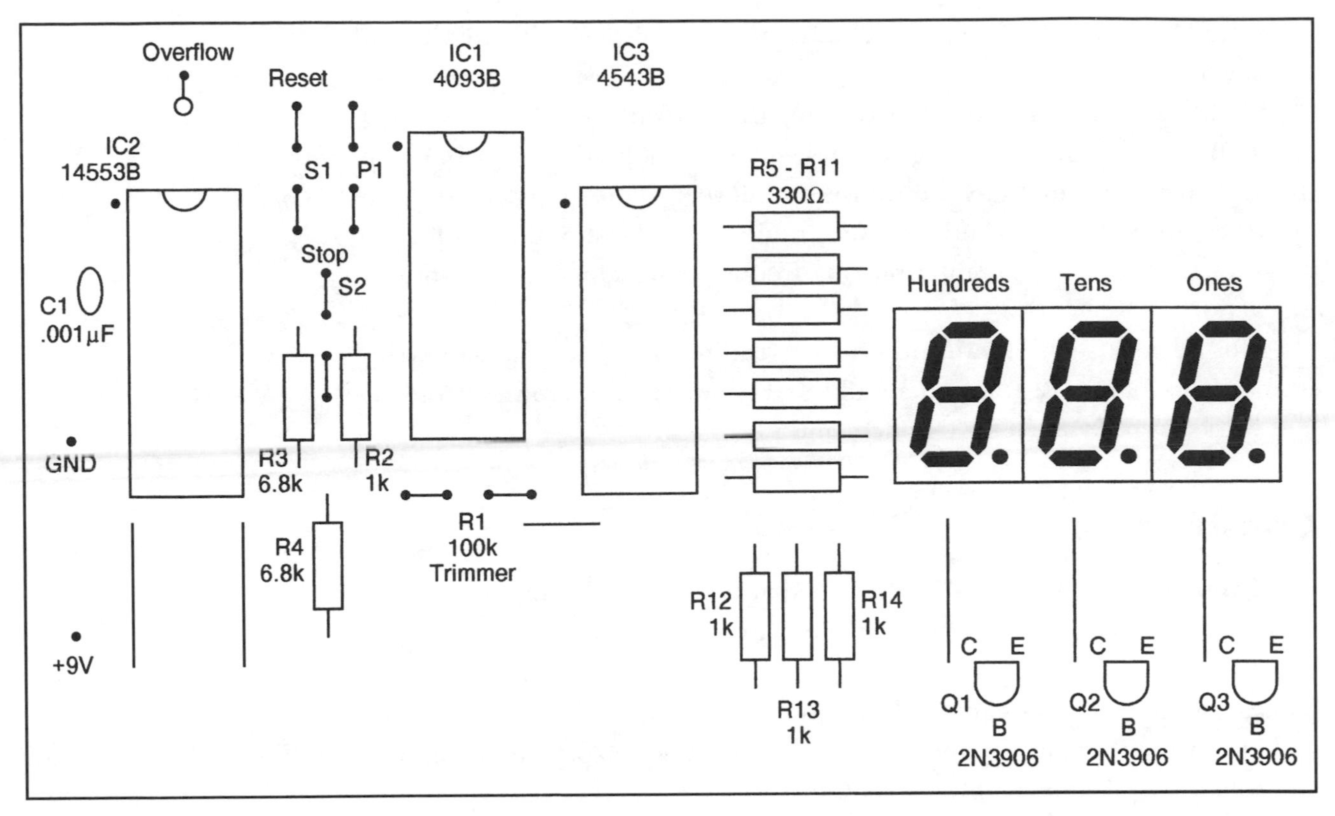

Figure 12-7: Parts layout and positioning for the photo-electric counter.

When wiring to the potentiometer, R1, you only need to connect to two of the three terminals, since R1 acts as a variable resistor in this circuit. One wire goes to the center terminal. Holding the potentiometer with the back facing you and the terminals at the bottom, connect the other wire to the right terminal. This will give you light maximum sensitivity when the potentiometer shaft, looking from the front, is rotated fully clockwise.

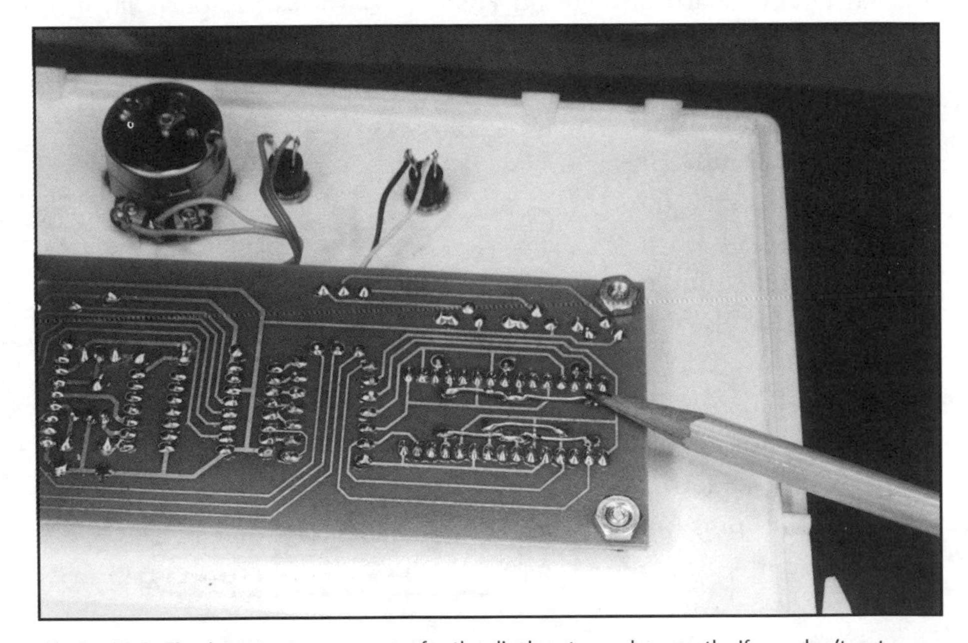

Figure 12-8: Five jumpers are necessary for the displays to work correctly. If you don't put these under the displays, you can put them on the etched side of the board.

Packaging

I decided to assemble the printed circuit board into a white plastic VHS video cassette tape case, as shown in Figure 12-9. I mounted everything on the inside of the case lid. I cut a rectangular hole for the displays, used four mounting screws to hold the PC board in place, and mounted the two switches and the potentiometer. The battery is held in position with double-sided tape.

The battery draws considerable current (about 55 milliamperes at 9 volts) so I added a simple slide switch to turn it off when not in use. This was more convenient than opening the case and unsnapping the battery connector. However, you may prefer to power the Counter externally, and this works fine. I found it worked well down to 5 volts (using only 20 milliamperes), and a regulated power supply was not required. A standard 6 or 9 volt DC wall plug type of power supply worked fine.

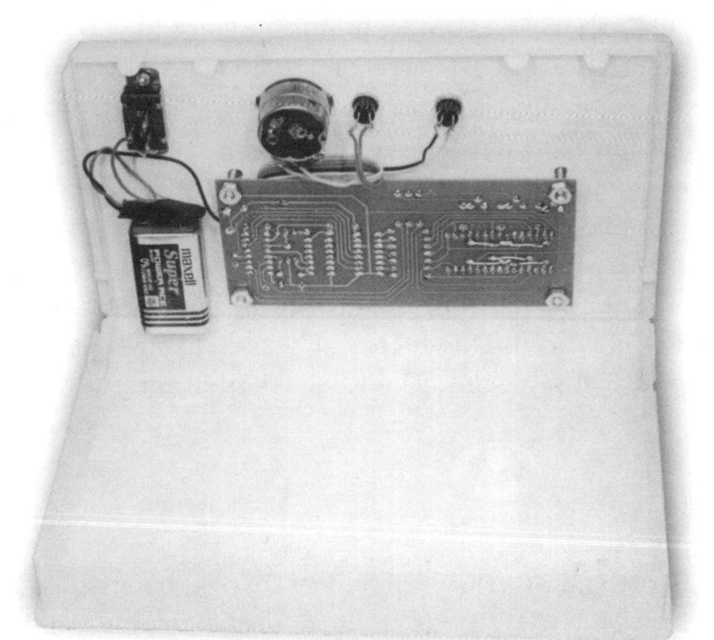

Figure 12-9: (left) An inside view of the circuit board and other components mounted in a VHS video cassette tape case.

Figure 12-10: (bottom) The completed photo-electric counter installed in a VHS cassette tape with labeling added.

Figure 12-10 shows the packaged photo-electric counter in the video cassette case after I added labeling. Looks terrific, doesn't it?

Testing

There's nothing much to testing the counter. Connect the battery or external power and all three digits should light, showing "O" for each digit. This tests all the LED segments except the center bar, G.

Dim the lights, or go in a closet, and use a flashlight for testing. The digits should increase one count each time the light shines on the photocell. Note that counting occurs NOT when the light beam is broken, but when the beam strikes the photocell again. It is easy to verify this. Shine a flashlight beam on the photocell, then cover the photocell with your finger (no count), and then uncover the photocell so the beam shines on it again (up one count).

The potentiometer has a definite effect on the counter's light sensitivity. Turn it clockwise to maximum sensitivity for a weak light source, counterclockwise to minimum for a strong light source. Watch the digits until they advance reliably after each resumption of the light path after interruption.

High ambient light can make counting unreliable since the photocell is very sensitive. It is best to use a light source aimed directly at the photocell. If ambient light remains too high, and interruption of the light source does not trigger a count when the interruption is removed, extend a one inch long black tube over the photocell to shield it from external ambient light.

The HOLD button may cause the digits to advance several counts whenever it is pressed. This is most common when the photocell is in complete darkness, or when the sensitivity control is turned to minimum sensitivity with some light on the photocell. This effect can be used to advance the digits manually to test all digit segments. The RESET button returns all digits to zero.

More Digits?

As mentioned earlier, an overflow condition (the next count after 999) causes pin 14 of IC2 to go HIGH. If this pin is connected to pins 1 and 2 of IC1 of an identical photo-electric counter (but without the P1 photocell!), and using a common power supply, you can count up to three additional digits (999,999). The first three digits will be the added counter. Since the added counter also has an overflow, you can add three more digits in the same manner with a third photo-electric counter, for a total count of 999,999,999!

Troubleshooting

Suppose, when you add power, none of the digits light! Check the wiring to the switches and potentiometer to see that you don't have them mixed up. Careful as I was, this happened to me! I moved the wires to the right place and some display segments lighted, but others didn't. This is when I discovered, by tracing circuit paths on the circuit board, and referring to the display pinouts shown in Figure 12-5, that I had omitted five jumpers that should have been put in place under the displays.

Of course, other things can go wrong. Be sure the transistors and ICs are oriented properly, and that the right values of resistors are where they should be. Voltage and ohmmeter checks can be useful. An oscilloscope can be used to see the multiplexing action on pins 2, 1, or 15 of IC2.

If any of the displays are not lighted, check the path from IC2 pins 2, 1, and 15 through transistors Q3, Q2, and Q1 to see if the multiplexing signal is switching between displays. To slow this action down so you can see the switching of each display, temporarily put a 0.22 µF capacitor across C1. If you use a 1 µF non-polarized capacitor across C1, you can see each display light separately for several seconds.

If any segment does not light for ALL three displays, trace the path for that segment back through its current dropping 330 ohm resistor to the proper output pin of IC3. Look for poor solder joints or solder bridges.

Generally speaking, unless badly handled or treated, the ICs, transistors or displays are most likely NOT at fault.

NOTE: *The above parts are included in the "CEKIT #K-031 Photo-Electric Counter Kit" available from Centerpointe Electronics, Inc., 5421 Lincoln Avenue, Unit A6, Cypress, CA 90360; telephone (800) 272-2737. Internet: www.shopsite.com/kits. The price at this writing was $24.95 plus $5 shipping, and 7.75% sales tax for California residents. The parts listed above as NTE are available from NTE Electronics, Inc. Call (800) 631-1520 for local distributor, price, and delivery. Pricing and availability of these items through Centerpointe and NTE may have changed since this book was published.*

PARTS LIST

C1:	0 .001 μF 50V ceramic disk capacitor
DISP1, DISP2, DISP3:	Common-anode 7-segment 0.56-inch high red LED display (NTE3078)
IC1:	CD4093B quad 2-input NAND Schmitt trigger (NTE4093B)
IC2:	MC14553B 3-digit BCD counter (NTE4553B)
IC3:	CD4543B BCD to 7-segment latch/ decoder/driver (NTE4543B)
P1:	Photocell (Radio Shack 276-1657 or equivalent)
Q1, Q2, Q3:	2N3906 PNP silicon transistor
R1:	100 kilohm linear potentiometer
R2, R12, R13, R14:	1 kilohm 1/4 W resistor
R3, R4:	6.8 kilohm 1/4 W resistor
R5 to R11:	330 Ω 1/4 W resistor
S1, S2:	Normally open pushbutton switch
Miscellaneous:	Etched/drilled/silkscreened printed circuit board, jumper wires, 9 volt battery snap, 1 14-pin IC socket, 2 16-pin IC sockets, solder.

CHAPTER 13

A Digital Thermometer

This digital thermometer can measure temperatures in all kinds of environments from -40 degrees to +100 degrees Centigrade. Since it's designed with a probe for dipping in liquids, it can also be used for ambient temperature measurements.

Many metals, semiconductors, and ceramics change their electrical resistance with temperature in a known and reproducible manner. This change in resistance results in a change of current or voltage, so it can be measured and then displayed in an analog or digital fashion. This is the basis for the digital thermometer described in this chapter.

This digital thermometer, which can be assembled from a kit or from scratch, can measure temperatures in all kinds of environments from –40 degrees to +100 degrees Centigrade. The temperature sensor can be mounted at the end of a cable or probe for dipping in liquids, or it can be used for ambient temperature measurements. For example, in repairing electronic equipment, the probe could be used to monitor the temperature of an electronic component suspected of overheating.

Three 0.56-inch high red LED 7-segment digital displays provide visibility in darkness. Although designed for short time portable battery use, simple changes allow use with plug-in external power that disconnects the battery. Although originally designed for measuring temperature in Centigrade (where 0 degrees is freezing, and 100 degrees is the boiling point of water), I'll describe the few changes to have it read in Fahrenheit (where 032 degrees is freezing and 212 degrees is the boiling point of water). I will also include temperature conversion formulas.

Some of the parts necessary for this project may be difficult to find or expensive. For example, I checked the price of the 40-pin ICL7107CPL 3-1/2-digit A/D converter integrated circuit from Thompson Consumer Electronics (their stock number SK10279), in 1997, and the cost was $25! (However, Mouser had it for under $7.) The three 7-segment displays cost about $2 each—if you could find them! And the LM335 temperature sensor integrated circuit may be available only from National Semiconductor distributors; I could not find a mail-order source or price. However, a kit of these and all other necessary parts is available from Centerpointe Electronics, as detailed at the end of this chapter.

Circuit Operation

For an overview of how this digital thermometer works, look at its block diagram, Figure 13-1. A known reference voltage is provided by IC4. This is amplified and adjusted by IC3 and fed to one input of the analog/digital converter, IC1. At the same time, the temperature sensor, IC5, feeds a voltage that depends on temperature to the other voltage input of IC1. IC1 is designed to compare the two voltage inputs and, using internal dual-slope conversion and drive circuitry, provides the appropriate outputs to light common-anode 7-segment LED displays.

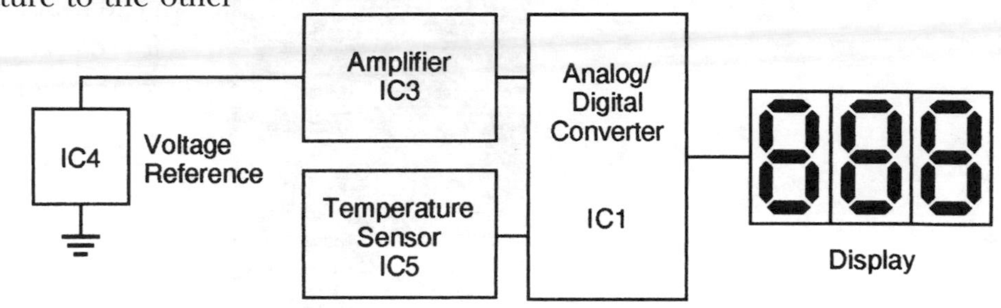

Figure 13-1: Block diagram of the digital thermometer.

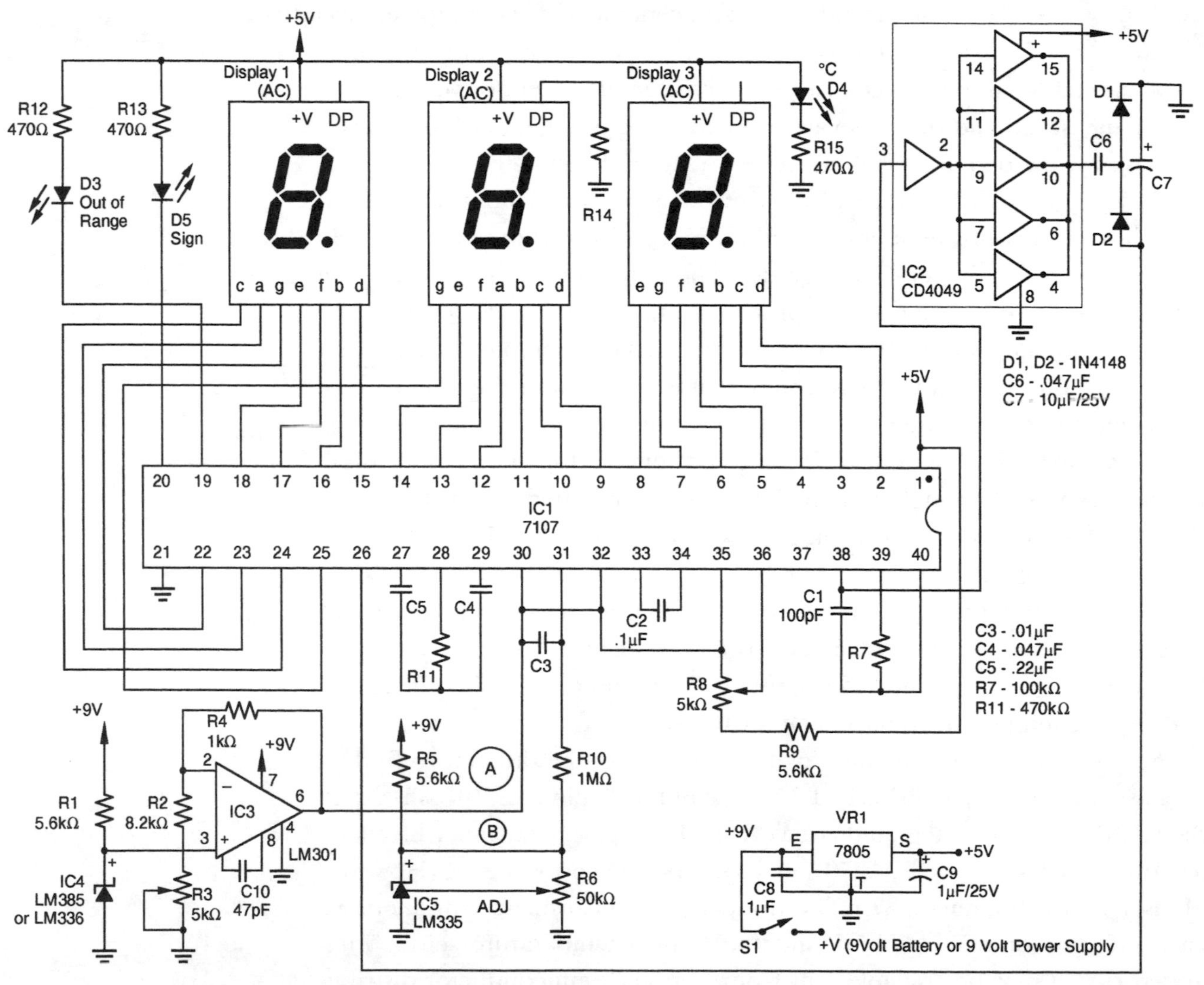

Figure 13-2: Schematic diagram for the digital thermometer.

The schematic, Figure 13-2, shows the complete circuit, and allows a more complete description of exactly how all this occurs. Obviously, the heart of the entire device is IC1, an ICL7107CPL 40-pin plastic dual-inline-pin (DIP) integrated circuit.

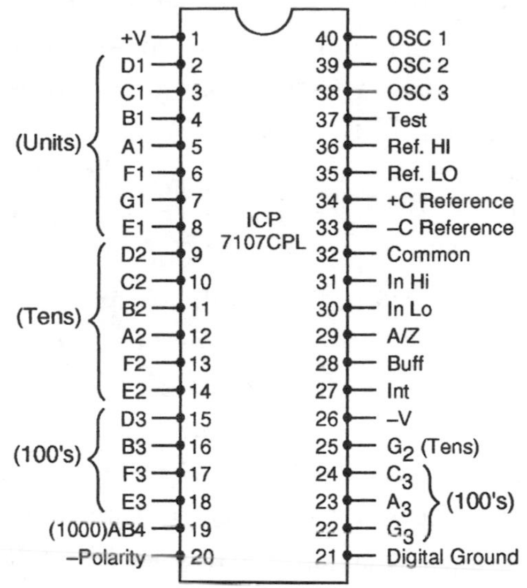

Figure 13-3: ICL7107CPL pinouts and functions.

The pinouts of IC1 are shown in Figure 13-3. IC1 is designed to operate from a +5V and –5V supply. However, to operate from a single +5V supply, a negative supply can be generated from OSC 3, pin 38, of IC1 with two diodes, two capacitors, and an inexpensive IC. As shown in Figure 13-2, IC2 (a 16-pin CD4049B hex inverter), C6, C7, D1, and D2 perform this function, and provide a DC voltage to pin 26, the V- pin of IC1.

Internal IC1 circuitry connected to pins 38, 39, and 40 is used with capacitor C1 and resistor R7 to create an oscillator operating at about 48 kHz as a "clock" used in the dual-slope conversion process. This provides about three readings per second. Capacitors C2, C3, C4, C5 and resistor R11 are external components required for proper IC1 operation. Pin 1 of IC1 receives a regulated +5V from a 9V DC battery or power supply feeding VR1, an LM7805 voltage regulator, together with filtering capacitors C8 and C9, with IC1 pin 21 as the ground return.

The voltage at pin 36, compared to the voltage at pin 35 of IC1, determines the scale at which IC1 measures input voltage. This is determined by the setting of multi-turn potentiometer R8, fed by resistor R9 from the regulated +5 volt supply, as you'll see later.

Two input voltages are provided to IC1 at pins 30 and 31. One is stable, the other varies with temperature, and IC1 measures this difference and converts this measurement to light the appropriate digital numbers on the display.

The voltage fed to IN LO, pin 30, of IC1 comes from the pin 6 output of IC3, an LM301 operational amplifier 8-pin integrated circuit. IC4, an LM336 2.5-volt Zener diode, is fed from the 9 V supply through resistor R1. Its Zener action provides 2.5 volts to pin 3 of IC3. However, the other input of IC3, pin 2 is adjusted by means of multi-turn potentiometer R3, R2, and feedback resistor R4, to provide a specific stable IC3 output voltage to pin 30 of IC1. I'll call this Voltage A. This voltage sets the temperature range to be read.

The voltage fed to IN HI, pin 31, of IC1 comes from a network composed of resistors R5 and R10, multi-turn potentiometer R6, and adjustable temperature sensor IC5, an LM335. IC5 is shown in Figure 13-4. This 3-pin temperature sensor looks like a transistor but is actually an integrated circuit with 16 transistors, two capacitors, and ten resistors! The voltage across its output is adjustable by the voltage seen at the ADJ pin. Once adjusted, the positive terminal then increases in 10 millivolt (0.01 volt) increments with each positive change of one degree Centigrade.

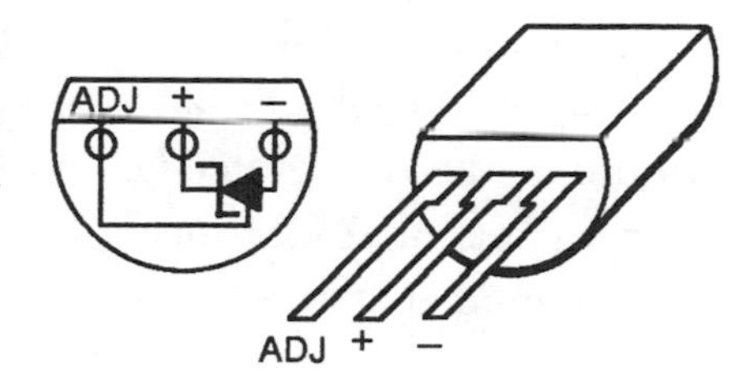

Figure 13-4: Pinouts of the LM335 temperature sensor IC.

The incoming 9 volts is adjusted by R6 so the voltage provided by the sensor at the intersection of R6 and R10 reflects the actual temperature of IC5. I'll call this intersection Voltage B. The difference between Voltage A (stable) and Voltage B (sensor temperature) provides the analog input to IC1, which then converts this analog voltage to a digital display readout.

Figure 13-5 is a pinout of the common-anode 7-segment LED display. Each display is connected to +5 V, with each segment of each display "seeking" ground. The output of IC1, by providing a path to ground, determines which segments of each display is lighted, and therefore the number displayed.

Resistor R14 provides a path to ground for the decimal point of the second display, so that the maximum displayed number is 99.9 . If that number is exceeded, pin 19 of IC1 provides a path to ground to red LED D3 and dropping resistor R12, lighting D3 for an "Out of Range" indication. If the display reads below 00.0, then polarity pin 20 of IC1 provides a ground path for yellow flat LED D5 and dropping resistor R13, lighting D5 to indicate a negative sign. LED D4 and dropping resistor R15 simulate a "degree" symbol.

Display Pinouts

1 - Cathode E
2 - Cathode D
3 - Common Anode
4 - Cathode C
5 - Cathode D.P.
6 - Cathode B
7 - Cathode A
8 - Common Anode
9 - Cathode F
10 - Cathode G

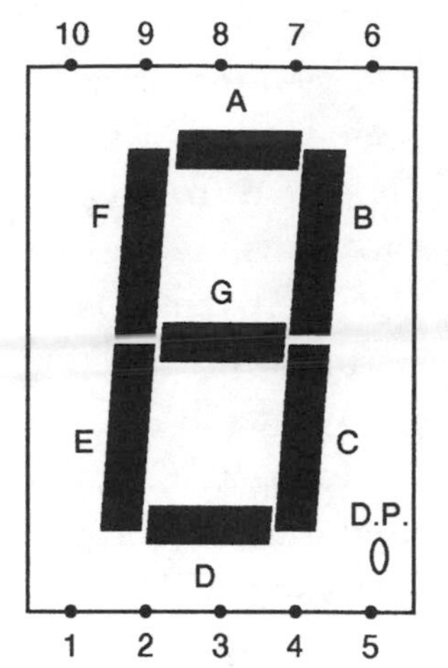

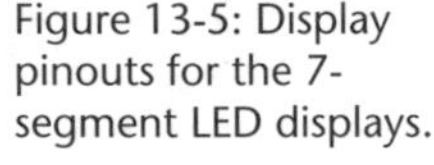
Figure 13-5: Display pinouts for the 7-segment LED displays.

Construction

As mentioned earlier, some of the parts required for this project may be difficult to find, or expensive. Although you could build this project from the information provided, I will assume you are assembling the digital thermometer using the parts kit.

The printed circuit board layout is shown in Figure 13-6, and the parts layout using this board is shown in Figure 13-7. A total of ten jumpers are used, and there are almost 200 solder connections!

Use special care in orienting polarity-sensitive parts (like diodes, LEDs, electrolytic capacitors) and properly orienting the integrated circuits in their sockets or solder locations. Also be certain to get the proper resistor

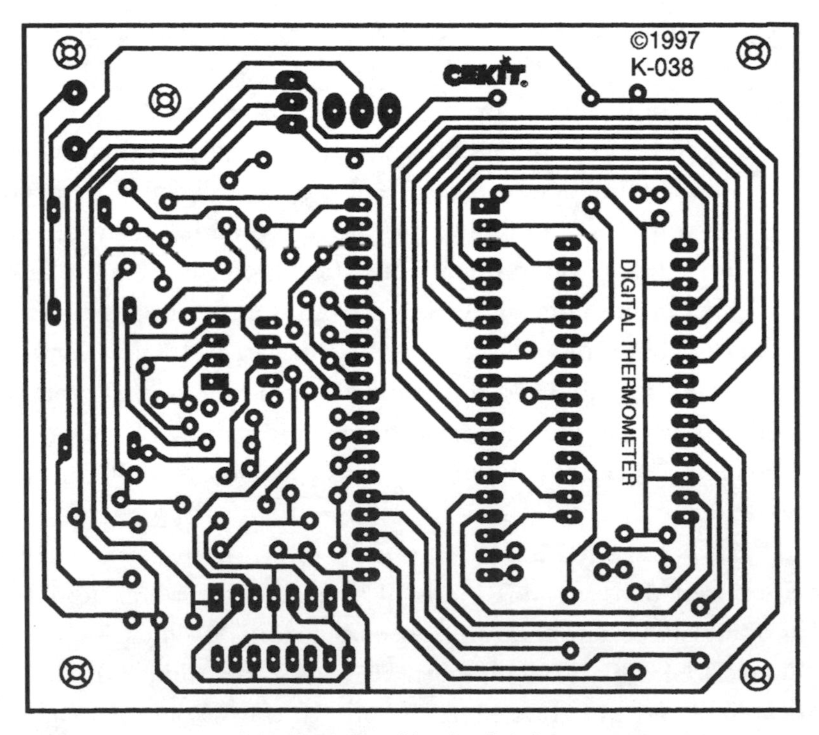

Figure 13-6: Pattern of the circuit board for the digital thermometer.

and potentiometer values in the proper locations or proper calibration will be impossible.

You must decide if you will be using your digital thermometer strictly as an environmental thermometer or to measure the temperature of fluids or other "off board" items. If you are going to measure off-board, you will need to make a temperature "probe" by connecting the three leads of IC5, the temperature sensor, through a three-lead cable to the proper pads on the printed circuit board. The sensor can be mounted sticking out of a simple tube (we used the front end of a plastic ball-point pen barrel). Fill the tube with epoxy for insulation between leads when placed in fluids.

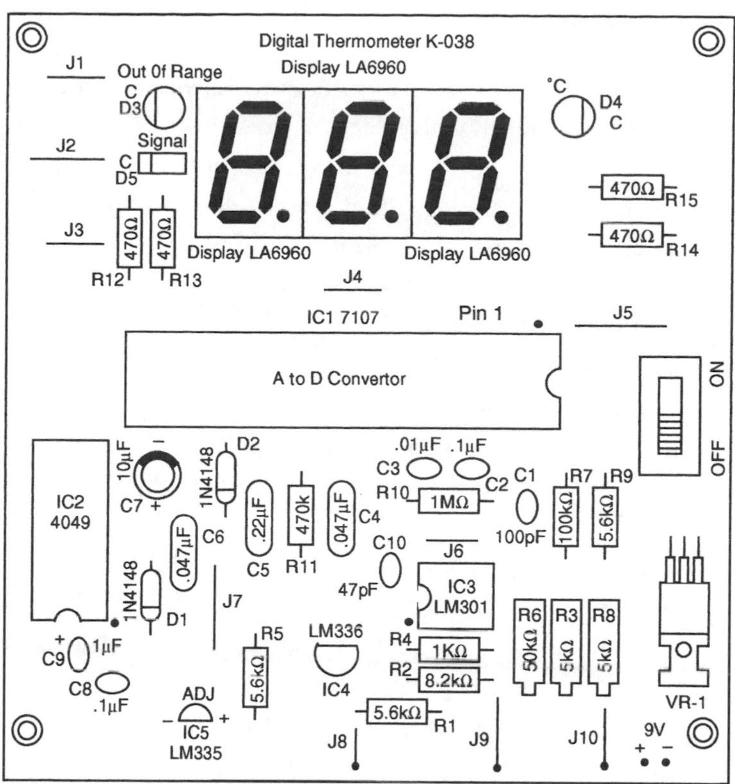

Figure 13-7: Parts layout on the circuit board for the digital thermometer.

You need to give thought during construction to how your digital thermometer will be packaged (discussed later). Depending on the method you use, you might wish to mount the power switch, S1, off the board. The voltage regulator, VR1, gets pretty hot in extended operation, and you might want to add a heat sink to the tab, bearing in mind that the tab is connected internally to the GROUND pin of VR1.

Figure 13-8 shows the circuit board for my digital thermometer before it was packaged. The wires you see go to the power switch, the battery snap, and the temperature sensor.

Figure 13-8: The circuit board of the assembled digital thermometer before being enclosed in a housing.

Quick Test

When a 9 V battery is connected, and switch S1 is turned on, all the displays and LED D4 should light up, as well as the decimal point of the center display. To make sure all segments of each of the three displays are connected, briefly connect +5 volts (NOT +9 volts!!) to TEST pin 37 of IC1. This should light all segments, showing 88.8 . (If not, refer to the "Troubleshooting" section of this chapter.)

Don't be alarmed if the reading does not seem to be correct, because the next step is to calibrate your digital thermometer.

Calibration

In order for your thermometer to have the proper range and "track" temperature properly, it is necessary to adjust the three potentiometers. R8 is adjusted so that exactly one volt, measured with a digital voltmeter, appears between pin 36 (positive voltmeter lead) and pin 35 (common voltmeter lead) of IC1. This sets the analog input voltage range of IC1 from 0 to 2 volts, allowing a range of 200 degrees at a sensor input of 0.01 volts per degree. This is beyond the range needed for reading –40.0 to +99.9 degrees Centigrade.

The next adjustment is made to R3, which is set to read exactly 2.73 volts between output pin 6 and ground pin 4 of IC3, the LM301 operational amplifier. This sets Voltage A (pin 30 of IC1) so that when Voltage A and Voltage B are identical, the display will read 00.0 .

The final adjustment is made to R6, and this depends on knowing the actual sensor temperature. If, for example, the sensor is at a room temperature of 22 degrees Centigrade, simply adjust R8 until the display reads 22.0 . If the sensor is in a liquid that has a known Centigrade temperature, use that figure for setting R6. In any case, the voltage at the intersection of R6 and R10 (Voltage B) will be higher or lower than pin 30 (Voltage A) by 0.01 volts per degree Centigrade difference between the sensor temperature and 0 degrees Centigrade.

The formulas for conversion between Fahrenheit and Centigrade are:

Fahrenheit degrees = (1.8 * Centigrade degrees) + 32

Centigrade degrees = (Fahrenheit degrees – 32) * 0.5555

Troubleshooting

If the displays don't come on when you apply power, check for approximately +9 V DC at the input pin of VR1 (right pin looking down on the circuit board with displays at the top of the board) and +5 V DC at the output (left) pin. The center pin and metal tab are ground.

Once you have power, the displays should come on. Connect +5 V briefly to TEST pin 37 of IC1 to assure that all segments (but not all decimal points) light. Looking at Figure 13-5, you'll see how the segments are identified (A - G). The schematic, Figure 13-2 (and Figure 13-3), show you which pins on IC1 drive each segment of each display. If a segment doesn't light during TEST, check the solder connections at IC1 and the display.

Any other problems can probably be traced to a bad solder joint, an incorrect part location, improper orientation of a polarity-sensitive part, or an IC plugged in "backwards." Troubleshooting is more of an art than a science, and takes patience as well as thought.

Modifications

As this was being written in 1997, the manufacturer of the parts kit was planning to modify the circuit board and provide the additional parts to allow you to build the kit for either Centigrade or Fahrenheit readout. If your kit has this option, follow the kit instructions.

However, if you are building from scratch with the circuit and printed circuit board shown, then there are only some small changes needed. For one thing, since your displays will now read in full digits (–032 to 212 and beyond), the decimal point is no longer needed, so remove resistor R14.

Resistors R16 (2.2K) and R17 (10K) are added to the circuit as shown in Figure 13-9. This is most easily done at the bottom of the printed circuit board.

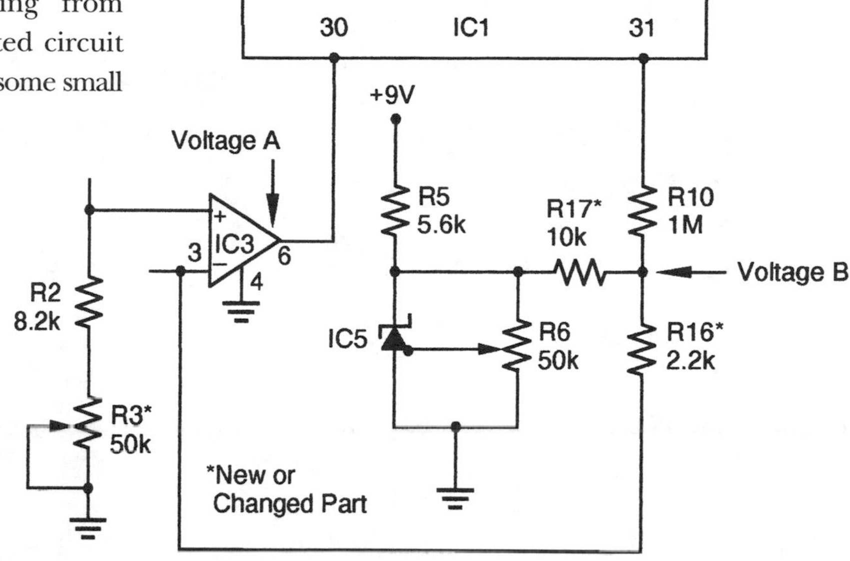

Figure 13-9: Schematic showing modifications to the digital thermometer.

Remove Jumper 7, noting the top and bottom solder pads from which it is removed. Now solder one end of R16 to the pin 3 pad of IC3. Solder one end of R17 to the bottom pad from which Jumper 7 was removed. Next connect the free ends of R16 and R17 to the top point where Jumper 7 was removed. Note that Voltage B is now at the intersection of R10, R16, and R17.

One other change is required. Unsolder 5K potentiometer R3 and replace it with a 50K multi-turn potentiometer. This completes the changes, but now you must recalibrate.

Set potentiometer R8, as before, to 1 volt between pins 36 (positive) and 35. This allows an analog input voltage range of 0 to 2 volts, the same as with Centigrade readings. Set the replaced potentiometer R3 to a new voltage of 2.554 volts from output pin 6 and ground pin 4 of IC3, the LM301 operational amplifier. This sets Voltage A (pin 30 of IC1) so that when Voltage A and Voltage B are identical, the display will read 000. With the new resistor network connected to IC5, the Voltage A to Voltage B difference is 1 millivolt (0.001 volts) for each one degree Fahrenheit change in sensor temperature.

Now adjust potentiometer R6 so the display reads the actual sensor Fahrenheit temperature, and you are finished with the conversion.

When operated on battery power, the digital thermometer uses about 200 milliamperes, which is a considerable drain for a typical 9V radio battery. As soon as the voltage drops below about 7 volts, the thermometer readings become erratic. If you wish to add the capability of powering the K-038 externally, any common wall-plug transformer DC power supply can be used, as long as it supplies about 9 volts at 250 milliamperes or more. Figure 13-10 shows how a standard closed-circuit phone jack can be wired to disconnect the battery when external power is used. If you use external power, be certain it is DC, and that the polarity of the external power is proper.

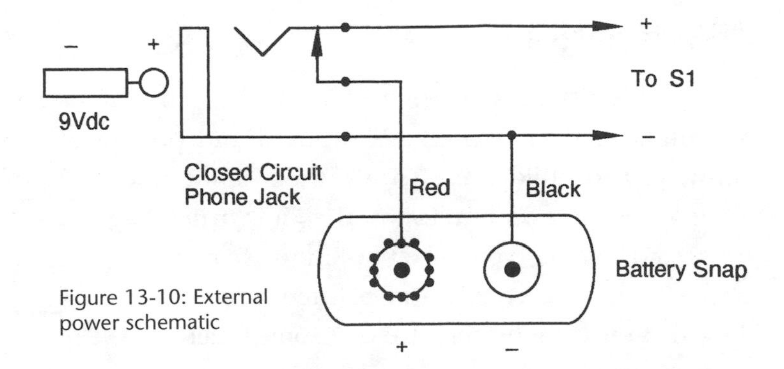

Figure 13-10: External power schematic

Packaging

As shown in Figure 13-11, I packaged my digital thermometer in an empty translucent case designed to hold a video cassette. Since it was made of thin plastic, it was relatively easy to cut access holes and a "window" to view the digital display.

The printed circuit board was mounted display-up in the cover of the case using four screws and nuts. The switch and external power jack were also mounted on the cover, and the battery was held inside the cover with double-sided tape. The sensor cable and probe come out through a notch in the bottom of the case. Labels were made using a desktop publishing computer program, and pasted to the cover.

Figure 13-11: The K-038 was packaged in an empty translucent plastic case from a video cassette, and labelled using a computer desktop publishing program.

A complete kit of all parts (possibly even including the optional parts by the time you read this) is available as the "CEKIT K-038 Digital Thermometer Kit" from Centerpointe Electronics, Inc., 5241 Lincoln Avenue, Cypress, CA 90630. The price at this writing was $34.90 plus $5 shipping and handling. California residents add 7.75% sales tax. Phones: (800) 272-2737 or (714) 821-1100; fax: (800) 493-7862 or (714) 828-0282. Pricing and availability may have changed since publication of this book.

PARTS LIST

(all resistors are 1/4 W 5% carbon)

R1, R5, R9:	5.6 kilohm
R2:	8.2 kilohm
R3, R8:	10-turn 5 kilohm potentiometers
R4:	1 kilohm
R6:	10-turn 50 kilohm potentiometer
R7:	100 kilohm
R10:	1 megohm
R11:	470 kilohm
R12, R13, R14, R15:	470 Ω

(all capacitors are 50 V ceramic disc unless otherwise specified)

C1:	100 pF
C2, C8:	0.1 μF
C3:	0.01 μF
C4, C6:	0.047 μF
C5:	0.22 μF
C7:	10 μF 25 V electrolytic
C9:	1 μF 25 V electrolytic or tantalum
C10:	47 pF
D1, D2:	1N4148 signal diode
D3, D4:	Red 5mm round LED
D5:	Yellow flat LED
IC1:	ICL7107CPL 3-1/2 digit A/D converter
IC2:	CD4049B hex inverter
IC3:	LM301 operational amplifier
IC4:	LM336 2.5 V Zener diode
IC5:	LM335 temperature sensor
VR1:	LM7805 5 V regulator
DISP1,2,3,4:	0.56-inch 7-segment high-efficiency red common-anode display (MAN6960 or equivalent)
S1:	SPST slide switch
Miscellaneous:	printed circuit board etched, drilled, and silk-screened; 9 v battery snap; three IC sockets (8-pin, 16-pin, and 40-pin); hookup wires; solder.

Optional parts for Fahrenheit conversion:

R16:	2.2 kilohm
R17:	10 kilohm
R3:	50 kilohm multi-turn potentiometer

CHAPTER 14

A Talking Alarm Clock

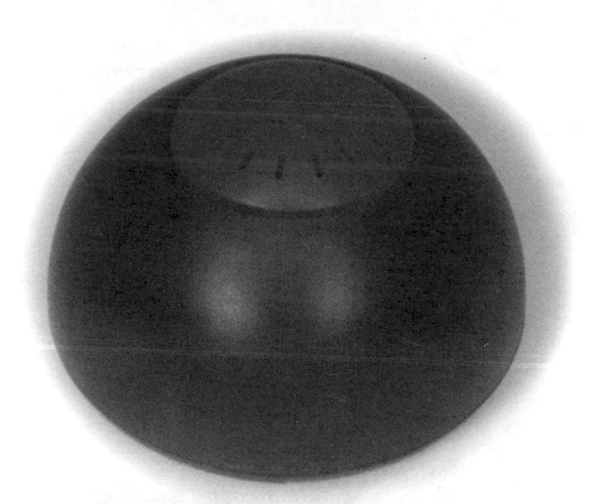

A talking alarm clock? Sure! This chapter describes one that gives "one touch" instant voice reporting of the hour and minute, and whether it is AM or PM. It can be set to announce hourly, or for any alarm time, with three different alarm sounds: beep, cuckoo, or rooster. A "snooze" function and volume adjustment are additional features.

Unlike other projects in this book, this one can't be made from "junk box" parts, or even parts you can buy independently. It is based on a custom integrated circuit, mounted onto a printed circuit board, that is only supplied with a parts kit. However, the use of custom ICs is a growing trend in electronics devices today, and that alone makes this project a good example of contemporary electronics technology.

Circuit Operation

All instruments used to measure the passage of time must have a time reference built into them. Figure 14-1 shows the schematic diagram of the AmeriKit AK-210 talking clock parts kit used to build this project. A tiny vibrating crystal is used as its time standard. This is shown as Y1 in the schematic.

Crystals are used in electronic oscillator circuits to provide a stable frequency of oscillation.

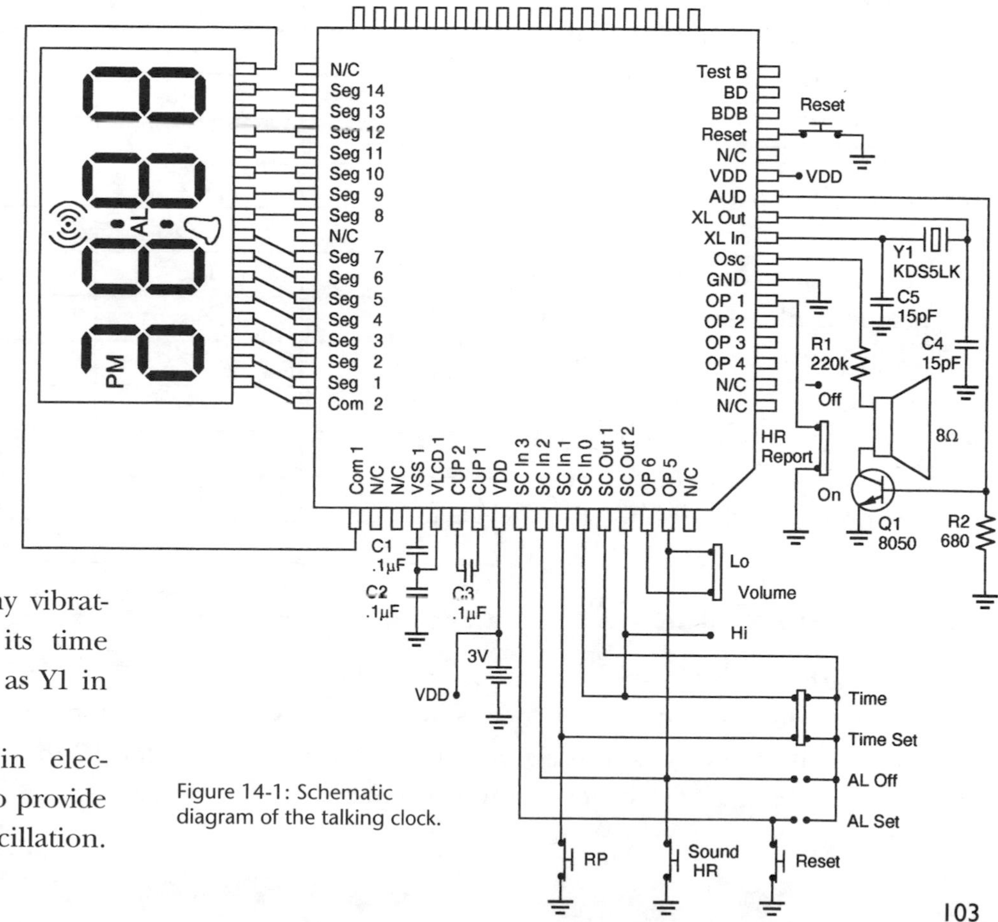

Figure 14-1: Schematic diagram of the talking clock.

A crystal can be thought of as the electronic equivalent of a mechanical tuning fork. When the tuning fork in Figure 14-2 is struck (excited) it will vibrate at only one frequency, producing a clear, undistorted tone. When the crystal in the clock circuit is excited (struck by an electronic pulse), it will also vibrate at only one frequency, as illustrated in Figure 14-3.

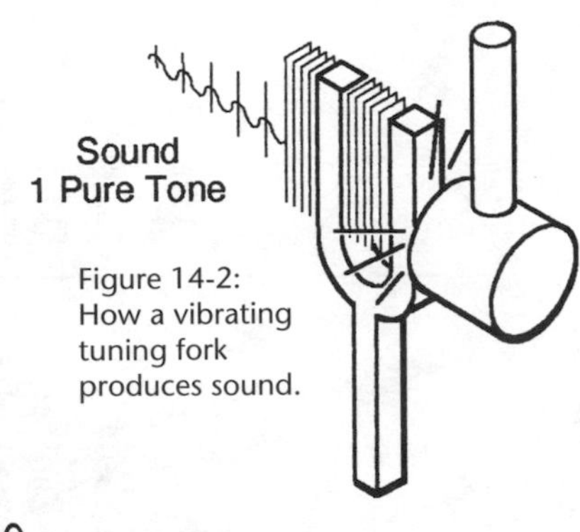

Figure 14-2: How a vibrating tuning fork produces sound.

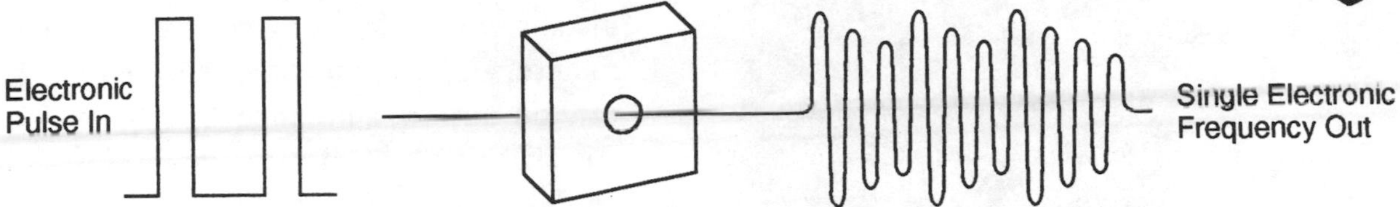

Figure 14-3: The crystal in the clock circuit will vibrate at only one frequency when excited.

The crystal supplied with the kit is designed to vibrate at a frequency of 32,768 Hz (Hz = Hertz = cycles per second). By continually dividing this frequency by 2, the frequency standard of 1.0000 Hz is obtained, as can be seen in Table 14-1.

Once the time standard of 1.000 seconds has been established, the custom integrated circuit attached to the AK-210 printed circuit board can calculate minutes, hours, and seconds. This information is then used to display digits or activate the appropriate speech.

This same built-on-board integrated circuit also checks to see if the time set in the alarm memory circuits has been reached. If the alarm time is the same as the time of day, the pre-selected alarm sound (beep, cuckoo, or rooster) will be activated. The alarm sound gets louder and repeats until it is deactivated or quits after 30 seconds.

NUMBER OF DIVISIONS	STARTING FREQUENCY	FREQUENCY DIVIDE BY	FINAL FREQUENCY
1	32,768	2	16,384
2	16,384	2	8,192
3	8,192	2	4,096
4	4,096	2	2,048
5	2,048	2	1,024
6	1,024	2	512
7	512	2	256
8	256	2	128
9	128	2	64
10	64	2	32
11	32	2	16
12	16	2	8
13	8	2	4
14	4	2	2
15	2	2	1

Table 14-1: By dividing the 32,768 Hertz crystal frequency by 2, fifteen times, the result is one cycle per second.

Making the Clock Talk

There are numerous ways to record and reproduce speech. Records, tapes, CDs, video, and digital memory circuits are just a few well known examples. This circuit uses words stored in digital memory to "speak" the time of day. Table 14-2 shows how many words to needed to verbally give the correct time.

The words in Table 14-2 are combined in certain ways to represent the correct time of day. For example "11:14 in the morning" would be "Eleven" "Four" "Teen" "_" "Oh" "Clock" "_" "A" "M." Note there are two pauses that separate words to improve the speech quality.

WORD	DIGIT OR PHRASE	WORD	DIGIT OR PHRASE
"Oh"	0	"Fifteen"	15
"One"	1	"Twenty"	20-29
"Two"	2	"Thirty"	30-39
"Three"	3	"Forty"	40-49
"Four"	4	"Fifty"	50-59
"Five"	5	"A"	A
"Six"	6	"P"	P
"Seven"	7	"m"	M
"Eight"	8	"_"	(Pause)
"Nine"	9	"Clock"	Clock
"Ten"	10	Alarm 1	(Sound of Rooster)
"Eleven"	11	Alarm 2	(Cuckoo Clock Sound)
"Twelve"	12	Alarm 3	(Beeping Sound)
"Thirteen"	13		
"Teen"	14,16,17,18,19		

Table 14-2: The words needed to "speak" the time are contained in the integrated circuit built onto the printed circuit board.

In practice these words are converted from a continuously changing voltage (analog) into a series of high (1) or low (0) voltage pulses (digital) by an analog to digital (A-to-D) converter. See Figure 14-4. When converting an analog signal to digital numbers the constantly changing waveform is sampled, and each sample is converted electronically to a digital format—in this case, a four-bit "word."

When only four bits are used, a decimal count of 0 to 15 is possible. Each bit has a "weight" based on the powers of 2, and only counts if it has a digital 1 value. 0 does not count. The leftmost of the four bits (the most significant digit) is 2 to the third power, or decimal "8." Moving to the right, the second digit is 2 to the second power, or decimal "4." The next bit to the right is 2 to the first power, or decimal "2." The rightmost bit (the least significant digit) is 2 to the zero power, or decimal "1." If all binary bits are 1 (on), you would add 8+4+2+1, for a total of decimal 15.

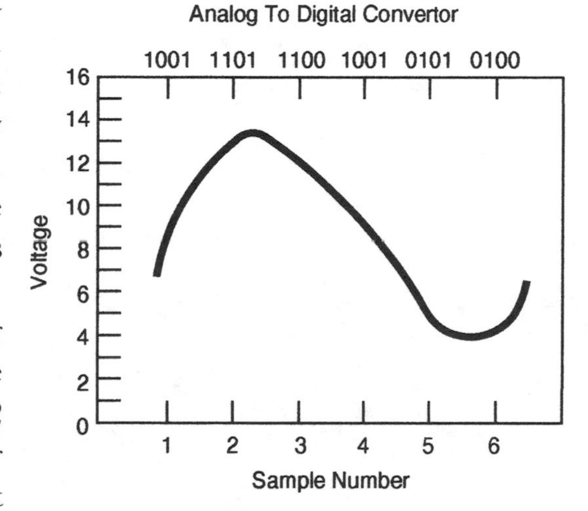

Sample Number	Voltage Analog	Digital
1	9	1001
2	13	1101
3	12	1100
4	9	1001
5	5	0101
6	4	0100

Figure 14-4: Analog to digital conversion.

Looking at Figure 14-4 you can see how the analog voltage sampled at nine different points each can be expressed as a four-bit digital word. For example, Sample #5 is 5 volts, which converts to digital 0101. Remembering that each digital 1 represents a power of 2, you evaluate from the left to the right and get (0+4+0+1=5.

The digital word samples are then stored in the integrated circuit and, when needed to operate the speaker, converted back to analog voltages electronically by a digital-to-analog (D-to-A) converter as shown in Figure 14-5.

If we say that a 1 means the voltage is present and a 0 means it is not, then 1001 would mean 8 volts is present, 4 volts is not, 2 volts is not, and 1 volt is present, yielding a total decimal value of 9.

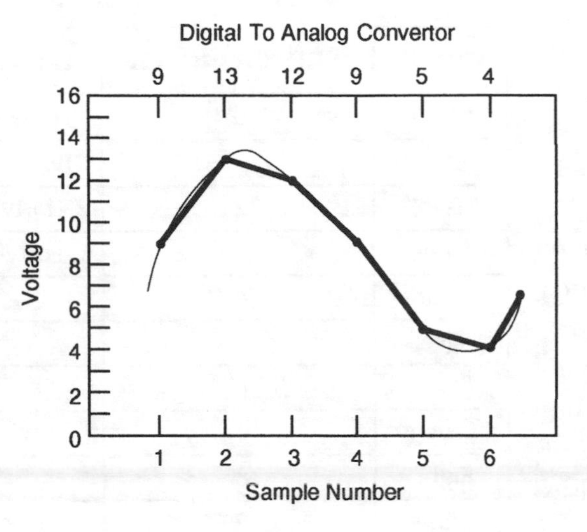

Digital	Analog
1001 = 1x8 + 0x4 + 0x2 + 1x1 =	9
1101 = 1x8 + 1x4 + 0x2 + 1x1 =	13
1100 = 1x8 + 1x4 + 0x2 + 0x1 =	12
1001 = 1x8 + 0x4 + 0x2 + 1x1 =	9
0101 = 0x8 + 1x4 + 0x2 + 1x1 =	5
0100 = 0x8 + 1x4 + 0x2 + 0x1 =	4

Figure 14-5: Digital to analog conversion.

After the digital numbers are converted to words, the words are amplified and sent to the speaker. The speaker reproduces the sounds that make the clock speak the time. As supplied, the AK-210 parts kit "speaks" only English. However, by special order, the integrated circuit used comes with words for any one of these foreign languages: Japanese, German, Spanish, Chinese, Italian, French, or Arabic.

The Visual Display

On the bottom of the AK-210 case is a small window where a four-digit time-of-day and some special symbols are shown using a liquid crystal display (LCD). Each number that appears in this window in made up of the appropriate combination of four vertical and three horizontal "segments," as shown in Figure 14-6.

Figure 14-6: How the LCD segments form various digits.

For example, the number "8" uses all seven segments. The number "3" uses only the two right vertical segments and all the horizontal segments.

Figure 14-7 shows a close-up of the display with ALL segments and symbols "on." Note that the leftmost digit does not have (or need) the upper left vertical segment, since it never needs to show other than a "1" or a "2" (for military 24-hour time). The kit as supplied uses only 12-hour AM/PM time. AM is not shown, but PM is. The symbol at the top center is to

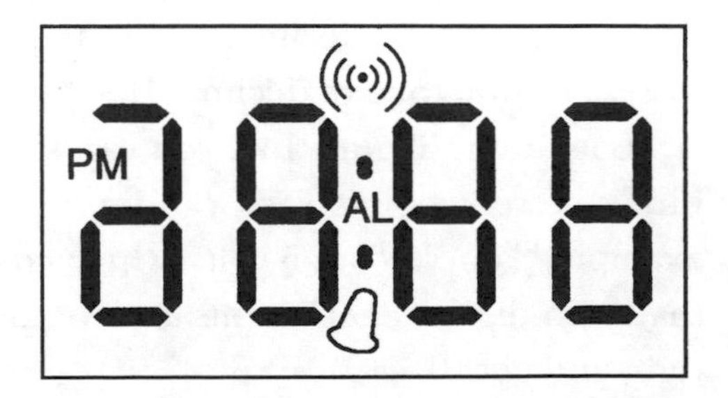

Figure 14-7: A view of the LCD display with all segments and symbols activated.

indicate the hourly announcement is active. The AL indicates the display is showing the alarm time. The little bell at the bottom center indicates the alarm is active.

Liquid Crystal Displays

The device used to create the black numbers and symbols on the bottom of the AK-210 is called a liquid crystal display. An organic material is placed between two glass plates in a pattern conforming to the segments and symbols desired to be displayed. The molecules in this organic material are normally disoriented, and ambient light reflects back to the viewer's eye, appearing as a metallic luster. The pattern is not visible.

However, as shown in Figure 14-8, when any of the pattern elements are exposed to an electric field (resulting from a voltage from the clock integrated circuit), the molecules of this element align themselves and allow ambient light to pass through to the back plate, where the light is absorbed. Since no light is reflected back to the viewer's eye, a black area appears conforming to the plated element pattern. This "black" area is much darker than the brighter metallic luster of the unaligned molecules in the unactivated area of the display.

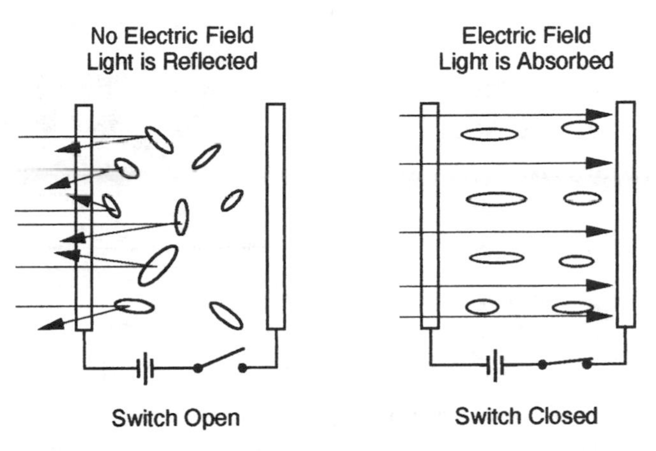

Figure 14-8: The internal operation of a liquid crystal display.

Construction

Figure 14-9 shows a top view of the completed unit. The dome in the center of the hemisphere is actually a switch. No other controls are visible.

An excellent, well-illustrated 16-page "Assembly and Instruction Manual" is included in the AK-210 parts kit. The kit includes wire and solder, as well as many special parts. The only things you'll need to supply (other than a soldering iron and pliers) are two "AA" 1.5-volt batteries. Each of the special molded or stamped parts is illustrated and identified in the manual, and each has its own part number if replacement is necessary.

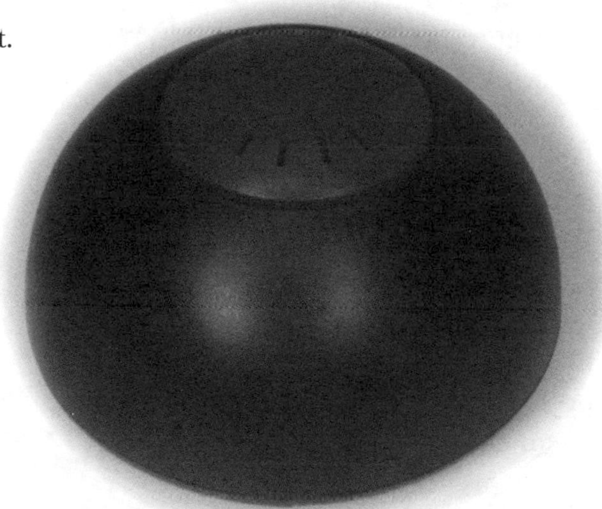
Figure 14-9: An outside view of the talking clock; touching the dome in the center lets you hear the time.

Assembly is straightforward. The only parts soldered to the printed circuit board (which is nicely silk-screened to show all part locations) are two resistors, five capacitors, one transistor, the crystal, three jumpers, and six wires that go to the battery compartment, speaker, and dome switch. All the wires are precut to length and ends stripped, making it even easier. And there was more than enough solder supplied!

The only difficulty encountered in construction involved the metal parts. In order to be installed properly, two of the four battery contacts had to be bent about 135 degrees instead of the 90 degrees shown in the assembly illustrations. Also, the contact fingers on all three switches had to be bent upward to make better contact with the printed circuit board after assembly.

Figure 14-10 shows the inside of the completed talking clock. Note how much room there is for all components and how simple the interconnection between circuit board, speaker, and power source is.

Figure 14-10: Internal view of the talking clock.

Testing

To assure proper operation before final assembly, the digital clock is tested by installing two AA batteries in the battery compartment to see if the LCD segment and symbols all function. If not, you might have to adjust the physical alignment of the LCD and its flexible multi-contact strip, called a "zebra."

Figure 14-11 shows a close-up of the switches and display on the bottom of the clock, while Figure 14-12 shows the entire bottom of the unit. With the LCD operating, you test out the various switch positions to see that you can set the present time and the alarm time. You also test the beep, cuckoo, and rooster alarm sounds, as well as the hourly announcement and volume settings. This is all adequately described in the manual.

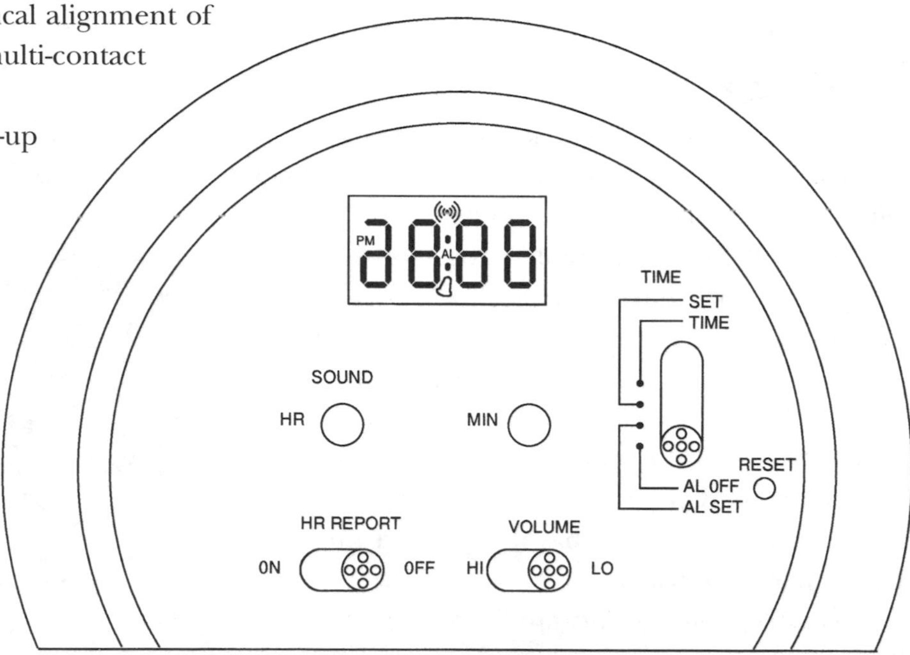

Figure 14-11: Close-up of the controls on the underside of the talking clock.

Once testing is satisfactory, the dome switch (a small circuit board contacted by pressing a rubber conductive cup) is installed, and the dome cover placed in position. You're done!

Figure 14-12: View of the underside of the talking clock.

Using

Simply tap the dome anytime and the clock will literally tell you the time. If the alarm is set, the sound you selected starts at alarm time and gets louder if you don't turn it off by moving a switch on the bottom. If you tap the dome switch instead, it acts as a "snooze alarm" and comes back on in 10 minutes.

If you set it for hourly announcements, it tells the time at the beginning of every hour. Setting the display for present time or alarm time involves setting a switch to the proper position and pressing an HOUR and MINUTE rubber button contact at the bottom of the clock to set the time. The RESET button sets it to 12:00AM.

The AmeriKit AK-210 Talking Alarm Clock Kit parts kit is available only from C & S Sales, Inc., 150 W. Carpenter Ave., Wheeling, IL, 60090.The kit price at the time this was written was $19.95. Add $5 North America shipping and handling. Illinois residents add 8% sales tax. For information or foreign orders call (847) 541-0743. North American orders: (800) 292-7711. Pricing and availability may have changed since this book was published.

CHAPTER 15

A DC Digital Voltmeter

This chapter teaches you about analog-to-digital conversion, 7-segment digital displays, and making voltage or current readings. And you'll end up with a portable digital voltmeter that will directly measure and display up to 19.9 volts DC with a resolution of 0.1 volts. The schematic, basic theory of operation, and a complete parts listing are provided.

There is hardly a handier piece of test equipment than a voltmeter. Equipment failures often result from a problem in the power supply. This can be anything from a weak or dead battery to a failed diode or transistor. The cardinal rule of troubleshooting is to first check the voltages.

Digital voltmeters are favored over older analog meters (meters with a needle that reads over a scale) by many electronics enthusiasts. There are several good reasons for this. Digital meters read numbers, meaning there is no "parallax error" caused by reading the meter scale from an angle. The decimal point is usually properly placed. Most digital meters also show a polarity sign in case you have the leads connected in reverse. Most importantly, the input resistance of most digital meters is somewhere around 10,000,000 ohms (10 megohms) regardless of the voltage being measured. This means digital voltmeters have virtually no effect on the circuit they are measuring.

The digital voltmeter described in this construction project has a 10 megohm input resistance, and is designed to read and display DC voltage from 0.1 to 19.9 volts in 0.1 volt increments. While this may seem like a limited range, most transistor, integrated circuit, and automotive voltages are below 20 volts.

No negative polarity indicator is displayed, but can be easily added. For convenience and portability, this project is battery operated, and four AA alkaline penlight batteries provide about 15 to 20 hours of continuous operation. Since no fragile moving needle movement is involved, this unit is relatively rugged. For those of you who like to experiment, later in this chapter I'll describe some changes you can make to add a polarity indicator and read higher voltages or DC currents.

7-Segment Displays

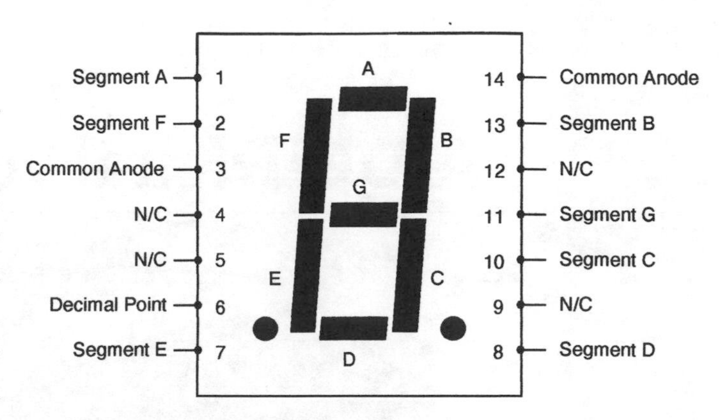

Figure 15-1: Pin connections for typical 7-segment LED displays.

Figure 15-1 shows the pin connections for the 7-segment LED displays used in this project. These are "common anode" displays, meaning each "segment" (lighted bar) is connected to positive voltage. Note that each segment has a letter designation—A, B, C, D, E, F, or G. In order for a segment to light, its pin must be connected to circuit ground, usually through a "driver" that consists of a resistor, or a transistor and its biasing resistors.

The combination of segments lighted at one time form a number. For example, light segments B and C form a number 1. Segments A, B, C, D, and G form the number 3. All segments together form the number 8. Get the idea?

These displays also appear to have a left and right decimal point. However, the actual displays used in this project have only the left decimal point connected internally.

Analog to Digital Converter

Figure 15-2 shows the heart of this design, the Teledyne TC807CPL $2^1/_2$ digit analog-to-digital converter. This 40-pin integrated circuit is designed to drive standard 7-segment LED displays without external drive electronics. That is, the resistors and transistors normally needed to light 7-segment displays are provided internally in this IC. This is a significant saving in parts. Although it would normally require 16 external transistors and up to 48 external resistors to light the 16 segments of $2^1/_2$ digits, they are not needed with the TC807CPL, since the necessary driver circuitry is built into the IC!

The operation of the TC807CPL is fairly complex, as reflected by the strange designations of some of the pins shown in Figure 15-2. Most of these functions will be explained—or at least touched upon—in the following section.

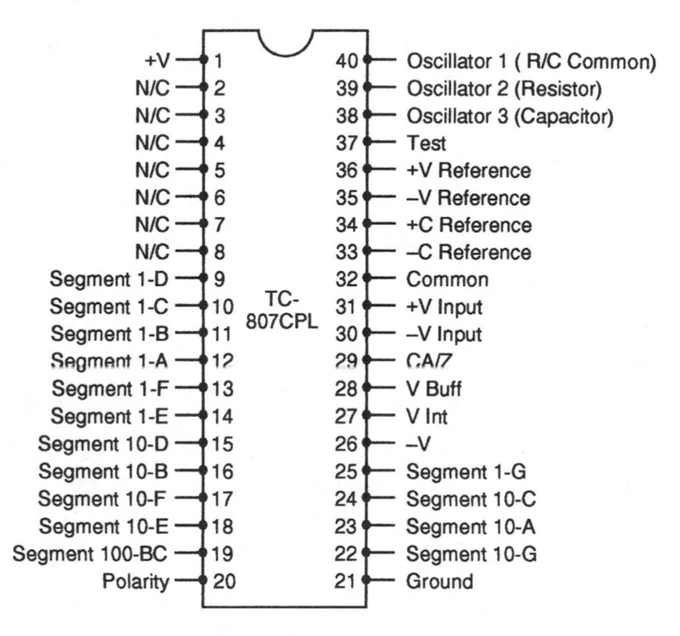

Figure 15-2: Pin connections for the TC807CPL IC.

Circuit Description

The schematic diagram of the $2^1/_2$ digit digital voltmeter is shown in Figure 15-3. U1, the TC807CPL, is a $2^1/_2$ digit analog-to-digital converter. The left digit can only count to 1, the others can each count from 0 to 9, hence the term “2 1/2 digit.” This IC contains all the necessary internal circuitry to process analog input voltages, convert them to digital signals, digitally count up to 199, latch and decode the count, and drive the three 7-segment LED displays.

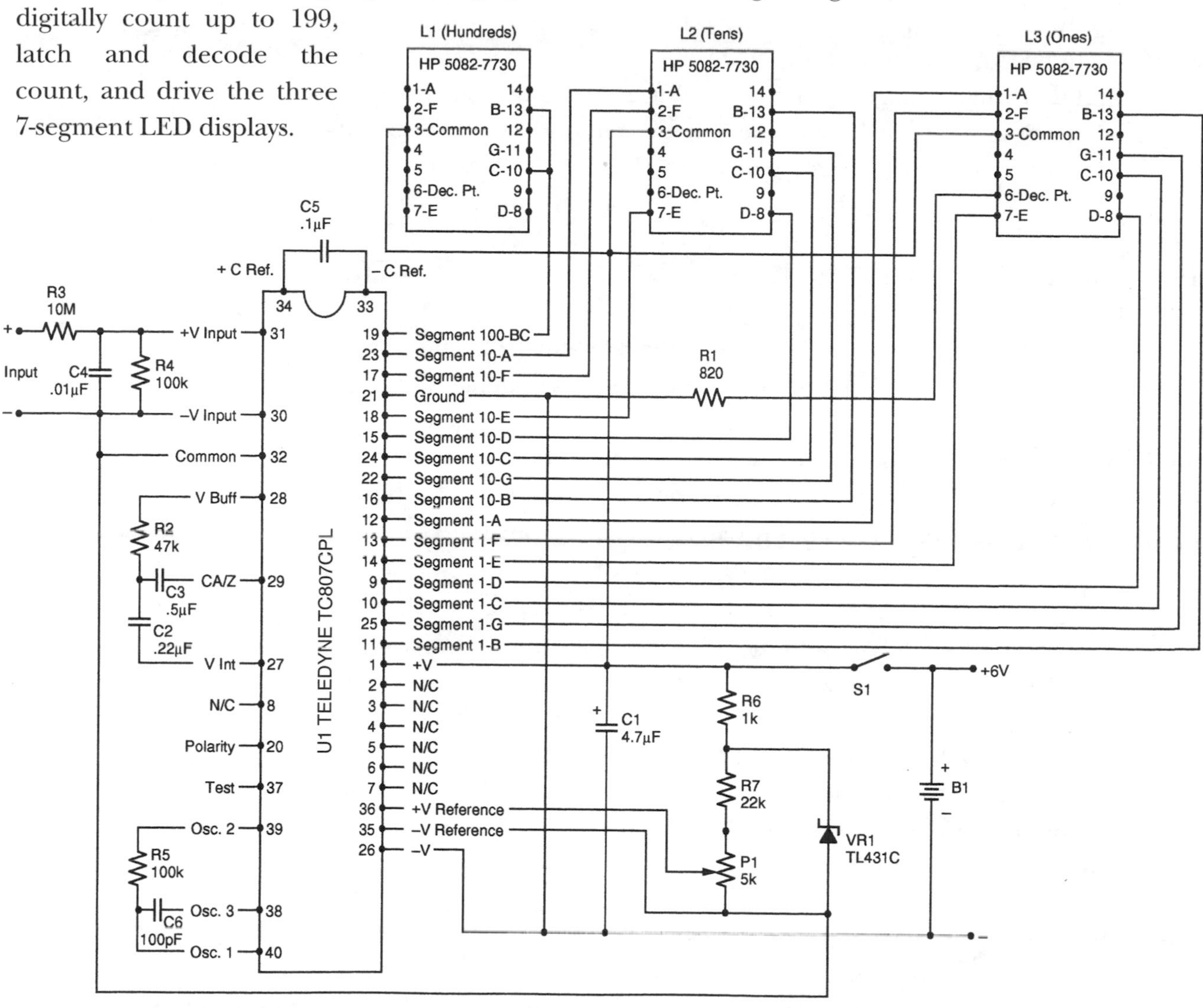

Figure 15-3: Schematic diagram of the digital voltmeter.

When switch S1 is closed, four AA alkaline batteries in a battery holder (B1) supply +6 volts to U1, the TC807CPL IC, at pin 1, and also the common anodes of the three 7-digit displays, L1–L3. The negative side of B1 is connected to pins 26 and 21 of U1, power ground. Capacitor C1 acts as a battery voltage filter.

VR1 is a voltage regulator that provides a stable reference voltage for the internal A-to-D converter at pins 36 (positive) and 35 (negative.) Resistor R6 is used to limit the current through VR1. Resistor R7 and potentiometer P1 form a voltage divider to set the calibrating voltage at pin 36 of U1. Notice that pin 35, the V- Reference voltage, is connected back to the V- Input, pin 30 and Common, pin 32. This insures that the input voltage to be measured is compared with the stable voltage-controlled reference rather than the diminishing power supply voltage.

The DC voltage to be measured is applied to the input, with the positive applied to resistor R3, and the negative to pin 30. R3 determines the meter's 10 megohm input resistance. Capacitor C4 filters out any AC noise which might be riding on the DC input voltage.

R3 and R4 form a voltage divider that reduces the input voltage applied to pin 31 down to the input operating voltage required by U1, 199 millivolts (0.199 volts) or less. The ratio of R4 to (R3+R4) is 100,000 divided by 10,100,000, or 0.0099, or about 0.01. That means if you are measuring 15 volts at the input, 15 times 0.01, or 0.15 volts (which is 150 millivolts) appears between pins 31 and 30. The meter would read 15.0 when properly calibrated.

How does this conversion from analog to digital happen? Each measurement cycle is divided into three phases: (1) auto-zero; (2) signal integration; and (3) reference integration. The conversion rate is set by the clock oscillator frequency, which is controlled by capacitor C6 and resistor R5, connected to U1 pins 38, 39 and 40. Using these components, the conversion routine measures and displays about three times each second.

The first phase takes place at pin 29, using auto-zero capacitor C3 and IC internal gates to establish a zero-input condition. Next, signal voltage integration takes place for a fixed number of clock cycles as a charge accumulates on capacitor C2 (pin 27) proportional to the input voltage. At the end of the signal integration period, reference integration is accomplished by having C2 discharge through buffer resistor R2 and pin 28. An internal digital timer measures the discharge time back to zero; the longer it takes, the higher the original input voltage.

The digital timer count is then decoded into three 7-segment LED drive signals through pins 9-19 and 22-25. In effect, each segment that is to be lighted is grounded through U1 with the appropriate internal driver circuitry.

The L3 decimal point (pin 6) is always ON, with its cathode grounded through current-limiting resistor R1 and the negative side of the power supply.

For proper reference in most applications, capacitor C5 is connected between U1 pins 33 and 34. Pin 20 provides a negative-polarity signal if the input voltage polarity is reversed; more on that later. Pin 37 provides a convenient test to see if all the display segments are correctly wired.

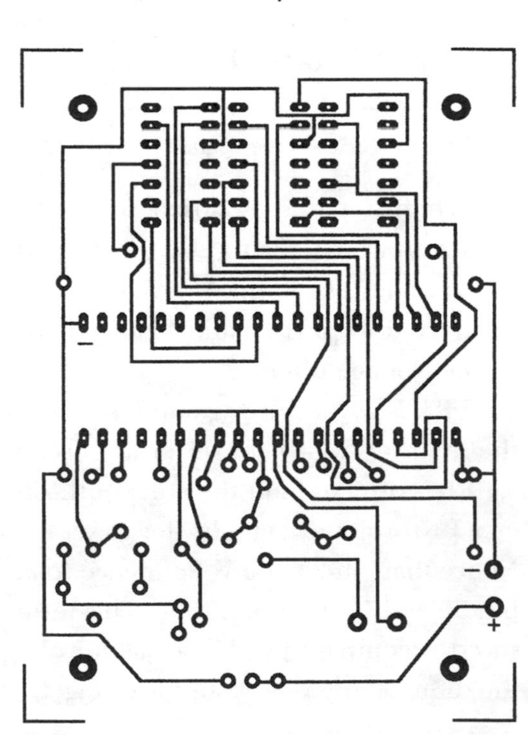

Figure 15-4: Printed circuit board pattern.

Construction

An etched and drilled printed circuit board is included in the parts kit described at the end of this chapter. Should you want to etch your own board, the board pattern is shown actual size in Figure 15-4. Some of the parts for this project may be hard to find in single-unit quantities, such as the TC807CLP and TL431C integrated circuits. If you already have most of the parts, but can't find some of the more "oddball" ones, there is a "partial" parts kit available. It is also described at the end of this chapter.

Fortunately, most of the parts used are common parts. However, the 7-digit displays deserve further discussion. There are many similar-looking 7-segment displays available, especially from dealers that sell surplus items. You can substitute for the specified Hewlett-Packard HP 5082-7730 displays ONLY if they are common anode displays with a left decimal point. If you use the printed circuit layout shown in Figure 15-4, you must also be sure the display pin connections are as shown in Figure 15-1, and that they will each fit in a 14-pin socket.

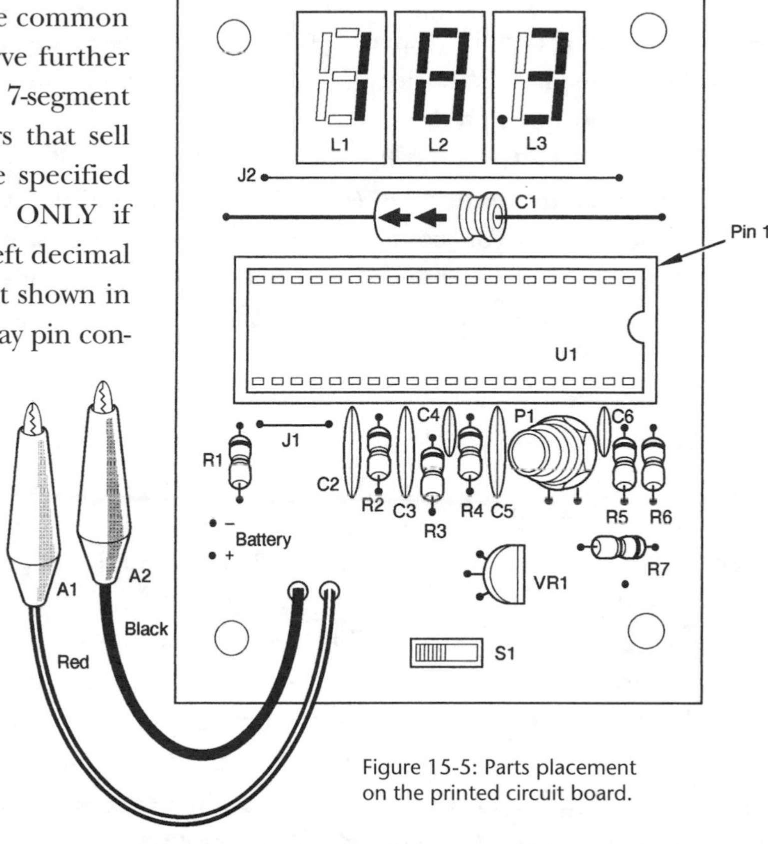

Figure 15-5: Parts placement on the printed circuit board.

Figurc 15-5 shows a pictorial of the assembly using the printed circuit board of Figure 15-4. Use a 25 or 35 watt soldering iron with a small tip to avoid overheating the components or having solder flow between pins or traces where it shouldn't be. First install and solder the resistors, then the capacitors (watch the polarity of C1), then the potentiometer, switch, voltage regulator (facing as shown), and sockets. Don't forget the two jumpers, using bare wire.

This is followed by soldering the battery holder wires to the board, and then the input lead wires, watching the polarity in both cases (red is positive, black is negative.) Now insert the displays and U1 into their sockets, with the display decimals at the bottom, and U1's notch on the right.

Figure 15-6 shows the completed circuit board resting in the case supplied with the parts kit. The 9 volt battery will give you a good idea of the size of the board. Note how close together the capacitors are; this is why careful soldering was stressed a few paragraphs ago.

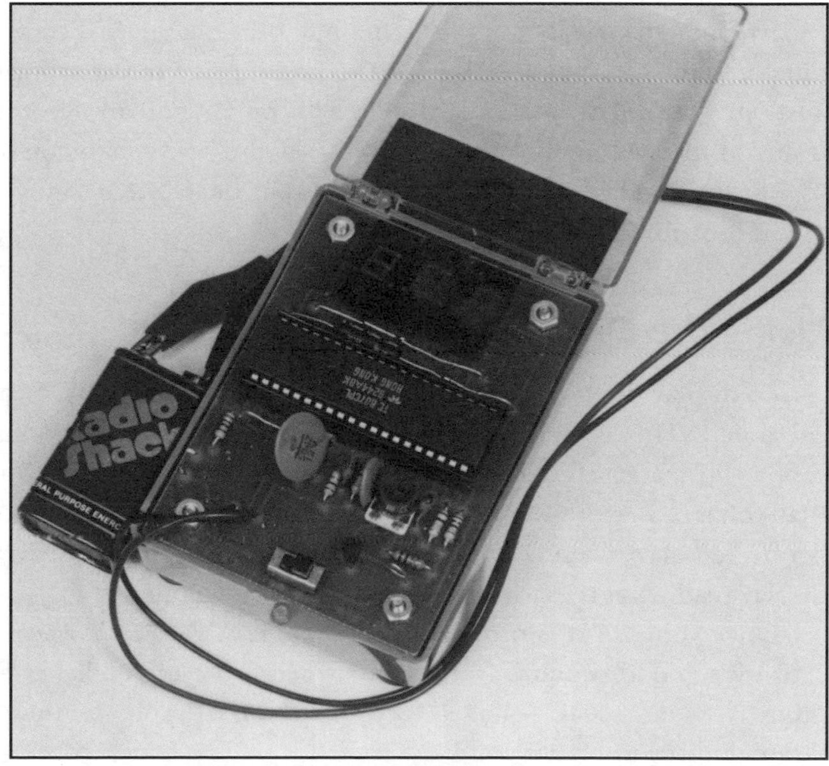

Figure 15-6: The assembled circuit board inside the case supplied with the parts kit.

Testing and Troubleshooting

Insert four AA alkaline batteries in the battery holder, making sure you observe battery polarity so the batteries are in series to provide a total of about 6 volts. Now turn on switch S1. After a short settling period during which L1 will probably display a "1" for a moment, displays L2 and L3 should each display a "0," with the left hand decimal point of L3 lighted. L1 should not now be lighted.

Now connect a fresh 9 volt radio battery to the input clip leads. The red clip lead should go to the solid (plus) 9 volt contact of the battery, the black lead to the other 9 volt contact. The display should now read somewhere around 9 volts. It may change from 8.9 to 9.0 or 9.1 (or other numbers, since the unit is not yet calibrated) as U1 samples the voltage about three times each second.

Now reverse the clip lead connections to the battery, intentionally connecting the polarity "backwards." The readings should be the same! In other words, this digital voltmeter is not polarity dependent.

But what if the digits don't light at all? In this case, check that your power supply batteries are connected properly and supplying about 6 volts. Be sure the polarity of the 6 volts as it connects to your circuit is not reversed.

Look to see if any solder bridges (pieces of solder crossing copper traces) have formed. If so, break or cut them with cutters, or with a desoldering iron, or using solder-absorbing copper braid. If you have a voltmeter, use the Figure 15-3 schematic to check for 6 volts wherever it should be. Be sure that U1 and the displays are plugged into their sockets in the right orientation, and that C1 and VR1 are not reversed.

If some display segments light, but others appear dead, connect a jumper wire from pin 1 of U1 (or the positive side of capacitor C1) to the Test pin 37 of U1. This should display a "1" on L1, and an "8" on L2 and L3. If any of these segments don't light, the Figure 15-3 schematic should lead you to the display segment pin and the pin on U1 that enables that segment. Probably a poor solder joint. . . .

Using the Digital Voltmeter

To display the proper voltage, your digital voltmeter (DVM) needs to be calibrated. This is simply a matter of connecting a known voltage to the input clip leads, and adjusting potentiometer P1 until the displays show that voltage. It is best if you have a known-accurate digital voltmeter also connected across the clip leads at the same time, and adjust P1 until your display reading agrees with the known-accurate voltmeter.

Once your DVM is properly calibrated, it will read voltages from 0.1 to 19.9 with good accuracy. Don't be alarmed if the digits flicker as the sampling is taken; some voltages vary by 0.1 or 0.2 volts as they are being sampled three times a second.

If you "overload" your DVM by connecting the input to a source greater than 1.99 volts, the last two digits (L2 and L3) will flash on and off, the over-voltage signal. I would avoid connecting the input to AC at all, or DC voltages above 30 volts. I don't know the limits, and I wasn't willing to destroy my DVM to find out!

It is important that you note that the V- power ground (pin 26) and the V- Input ground (pin 30) are NOT the same. In other words, the input voltage to be measured is compared with the reference voltage, NOT the power supply voltage. For this reason, you can't measure its own power supply with this meter! If you try, you won't hurt anything; you'll just see a decimal point on L3, and a "1" displayed on L1, with L2 and the other segments of L3 apparently dead! This is also the reason you can't use this meter, without modification, to read car battery voltage if it's powered by the car battery.

Packaging

It is recommended that the DVM be mounted in a case. A metal, wood, or plastic case is usually used for this purpose. If you use a metal case, be sure none of the circuitry is in contact with the case. The case supplied with the parts kit makes a very attractive finished unit, as shown in Figure 15-7.

Figure 15-7: The digital voltmeter inside the packaging supplied with the parts kit.

Modifications

The easiest modification is to add a polarity indicator. Just take a standard jumbo LED and connect the anode to positive 6 volt power, and the cathode (usually the lead at the flat spot on the LED base) to pin 20 of U1. Now when you reverse the polarity of the input voltage, this LED will light. No current-limiting resistor is necessary. You could also use segment G of L1 (not presently connected) for this purpose; just connect pin 11 of L1 to pin 20 of U1.

If you'd like to read higher DC voltages, it is only necessary to change the relationship of input voltage divider resistors R3 and R4. The smaller R4, the greater the input voltage before exceeding the 199 millivolt maximum between pins 31 and 30 of U1. Just for fun, temporarily connect another 100 K resistor directly across R4. This effectively makes the resistance of this leg of the divider about half of what it was. This should also cut the input voltage reading in half; 9 volts will now read 4.5 volts. Just double the

DVM reading. However, keep in mind my earlier warnings about not using this to measure DC voltages over 30 volts or any AC voltages whatsoever!

You can also use your DVM to read DC current flowing in a circuit. You do this by putting a known accurate resistance in series with the circuit you wish to measure, and then using the DVM to measure the voltage drop across the known resistor. Using Ohm's Law (E/R=I), you then know that the measured DVM voltage divided by the resistance value equals the current flowing in the circuit.

Don't expect this DVM to compete with the flexibility and accuracy of more expensive units. It is entirely practical and useful for measuring DC voltages below 20 volts, and lends itself very easily to experimentation from which you learn and have fun at the same time.

NOTE: *The following items are available from Cal West Supply, 31320 Via Colinas, Suite 105, Westlake Village, CA 91362. Phones: (800) 892-8000 or (818) 889-2209. As this is written, the minimum order is $10, not including shipping, and all orders in the Continental U.S.A must include $3.50 shipping and handling per order. California residents must add sales tax. Alaska, Hawaii and Canada must include 20% (minimum $6) for shipping and handling.*

HK621-Complete parts kit: all items listed above, including printed circuit board, plastic case, and mounting hardware, $16.65.

HK621SP-Special parts kit: printed circuit board, U1, 40-pin socket, VR1, P1, S1, $11.70.

Pricing and availability of these items may have changed since publication of this book.

PARTS LIST

C1:	4.7 μF 35 WVDC electrolytic capacitor
C2:	0.22 μF ceramic disc capacitor
C3:	0.5 μF ceramic disk capacitor
C4:	0.01 μF ceramic disc capacitor
C5:	0.1 μF ceramic disc capacitor
C6:	100 pf ceramic disk capacitor
L1, L2, L3:	Hewlett-Packard 5082-7730 0.3-inch 7-segment LED display, or equivalent (see text)
P1:	5 kilohm 1/2 W potentiometer
R1:	820 Ω 1/4 W 10% carbon resistor
R2:	47 kilohm 1/4 W 10% carbon resistor
R3:	10 megohm 1/4 W 10% carbon resistor
R4, R5:	100 kilohm 1/4 W 10% carbon resistor
R6:	1 kilohm 1/4 W 10% carbon resistor
R7:	22 kilohm 1/4 W 10% carbon resistor
S1:	Single-pole single-throw (SPST) slide switch.
U1:	Teledyne TC807CPL 2-1/2 Digit analog-to-digital converter IC (see text)
VR1:	Texas Instruments TL431C voltage regulator IC
Miscellaneous:	3 14-pin sockets, 40-pin socket, 4 AA-cell battery holder, alligator clips, wire, solder. Optional: printed circuit board, plastic case, mounting hardware.

CHAPTER 16

A Digital Multimeter Kit

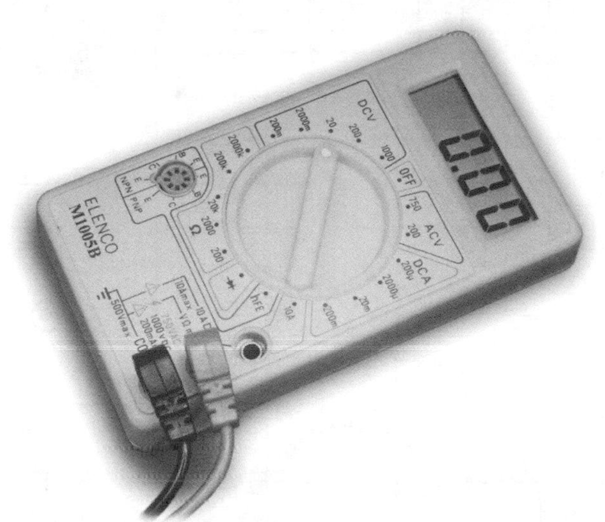

One of the best ways to learn how a digital multimeter (DMM) works is to build your own. The best way to build your own is to use a parts kit, since many of the necessary parts are hard to find (and expensive!) individually. The unit described in this chapter is useful for voltage, current, and resistance measurements as well as diode and transistor testing.

The digital multimeter (DMM) has become the most commonly used test instrument in electronics. While low-end DMMs measure AC and DC voltage, DC current, and resistance, many other functions are common in higher-priced DMMs. For example, it is not uncommon for a DMM to measure transistor gain and to test diodes and continuity.

Since the digital conversion circuitry and display are already in a DMM, some more elaborate units include capacitor, inductor and frequency measurement as well. It is unusual, however, to find a $3^1/_2$ digit DMM with transistor and diode tests in addition to the more conventional voltage, current, and resistance measurements, at a low price. The Elenco M-1005K parts kit will let you build such a unit. Because many of the parts used in building this DMM are custom made, it is not possible to build it from your "junk box" or by using parts from other sources.

By assembling this kit, you can learn how the heart of a typical DMM—the analog to digital (A/D) converter—works. Also, since my assembled kit initially didn't operate properly, I learned how to troubleshoot the popular 7106 $3^1/_2$ digit LCD DMM integrated circuit chip with an oscilloscope.

Since the Elenco M-1005K parts kit comes with a detailed, illustrated assembly manual, this chapter will describe the theory of operation and applications of this DMM. Details on obtaining the parts kit are contained at the end of this chapter.

Circuit Description

Figure 16-1 shows a complete schematic of the digital multimeter. This is difficult to follow because of all the resistors and switching circuits, but begins to make sense after some thought. The buss bars, drawn here horizontally, are etched on the actual circuit board in concentric circles to mate with six metal sliding contacts positioned under the rotary switch knob.

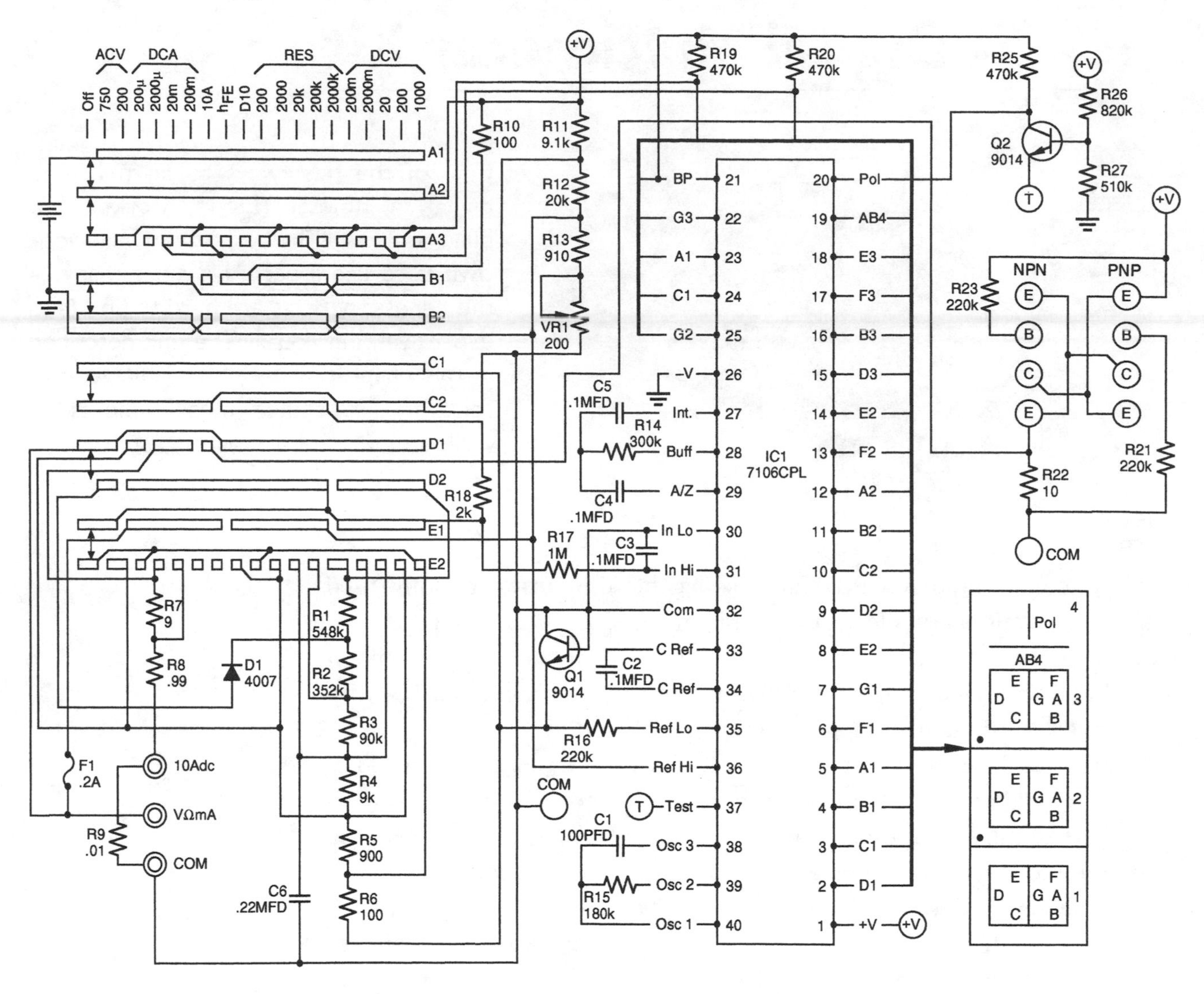

Figure 16-1: Digital multimeter schematic diagram.

Refer to Figure 16-1. If you visualize how the double-arrow contacts in the OFF position move to the right one position at a time as the rotary selector knob is turned clockwise, you can actually draw an individual circuit for each of the 20 switch positions. This could make a neat class project for 20 students, each assigned the task of drawing the schematic for one of the positions.

Theory of Operation

A block diagram of the DMM is shown in Figure 16-2. The input voltage or current signals are conditioned by the selector switches to produce an output DC voltage with a magnitude between 0 and 199 millivolts (mV). If the input signal is 100 V DC, for example, it is reduced to 100 mV DC by selecting a 1000:1 voltage divider circuit. Should the input be AC, it is first rectified and then divided down.

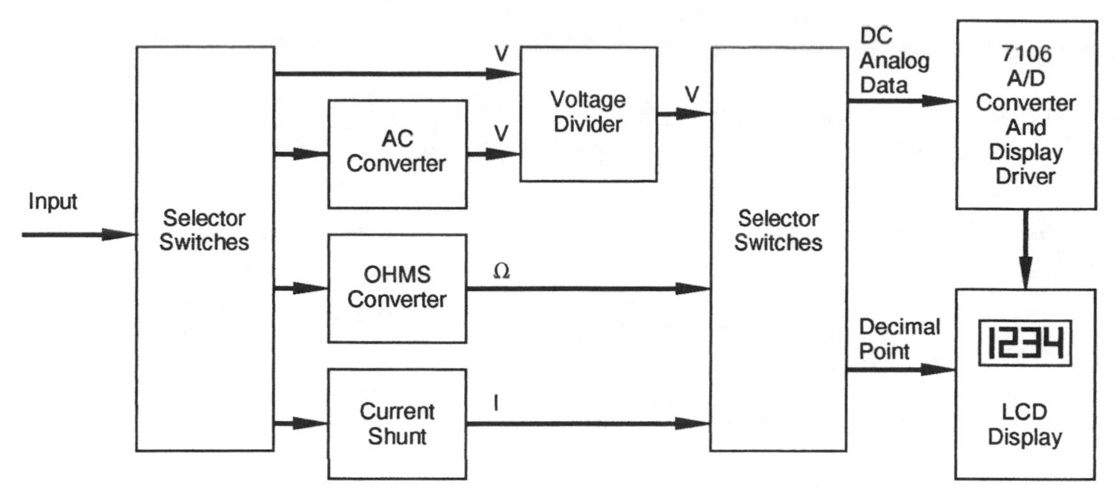

Figure 16-2: Block diagram of the DMM.

For resistance measurements, an internal voltage source drives the test resistor in series with a known resistor. The ratio of the test resistor voltage to the known resistor voltage is used to determine the value of the test resistor. If current is to be read, it is converted to a DC voltage by internal shunt resistors.

The resulting DC analog data is fed to a very popular integrated circuit, the 7106, which performs the functions of converting the analog data to digital, and also provides the seven-segment decoder and display drivers for the LCD display.

NOTE: *The 7106 IC comes in various physical forms. The 40-pin dual-inline pin (DIP) version used here, and found in many other inexpensive 3-1/2 digit LCD DMMs, is the 7106CPL.*

If you try to buy this IC separately from any of the several manufacturers, it costs almost as much as this entire kit! Some DMMs use the 7106R, with the "R" indicating that the pinouts are a reversed left-right mirror-image of the plain 7106. That is, pin 1 would be pin 40.

This kit uses the regular 7106.

The pinout of the 7106 is shown schematically in Figure 16-1. The voltage input to the 7106 IC, pin 31, is fed to an internal analog-to-digital (A/D) converter. Here the DC voltage is changed to a digital format. The resulting signals are processed in the decoders and appear on pins 2-25 to light the appropriate LCD backplane (pin 21) and segments.

Timing for the overall operation of the A/D converter is derived from an external oscillator (pins 38, 39 and 40). The external resistor R15 and capacitor C1 values result in an oscillator frequency of around 25 kHz. In the IC, this frequency is divided by four to 6250 Hz, and fed to dual-slope integrated A/D circuitry, explained shortly, that provides an LCD readout about twice per second. The digitized measurements are presented to the display as four decoded seven-segment digits, plus polarity. The decimal point position on the display is determined by the selector switch setting.

A/D Converter

A simplified circuit diagram of the analog portion of the analog to digital (A/D) converter is shown in Figure 16-3A. Each of the switches shown represents analog gates which are operated by the digital section of the A/D converter. The basic timing for switch operation is keyed by the external oscillator. The conversion process is continuously repeated.

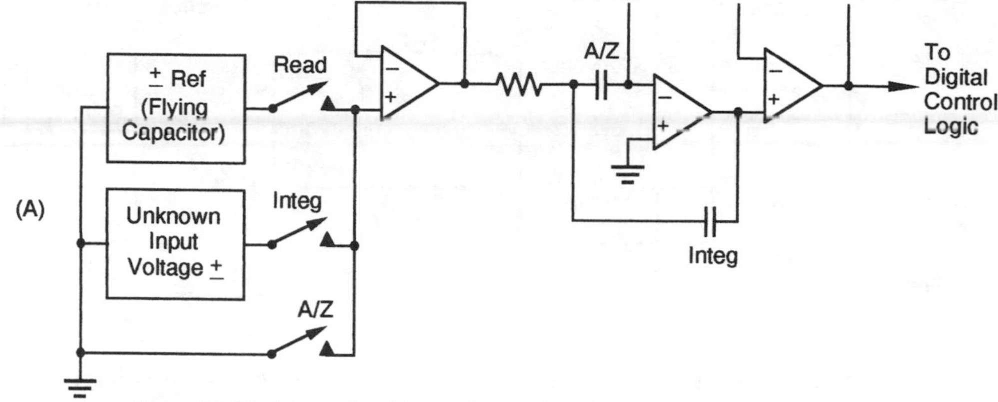

Figure 16-3A: Schematic of the analog section of the A/D converter.

A complete A/D conversion cycle is shown in Figure 16-3B. A measurement cycle can be divided into three consecutive time periods: auto-zero (AZ), integrate (INTEG), and READ. The 6250 Hz counter determines the length of each count as 160 microseconds (1/6250), or 160 milliseconds for 1000 counts.

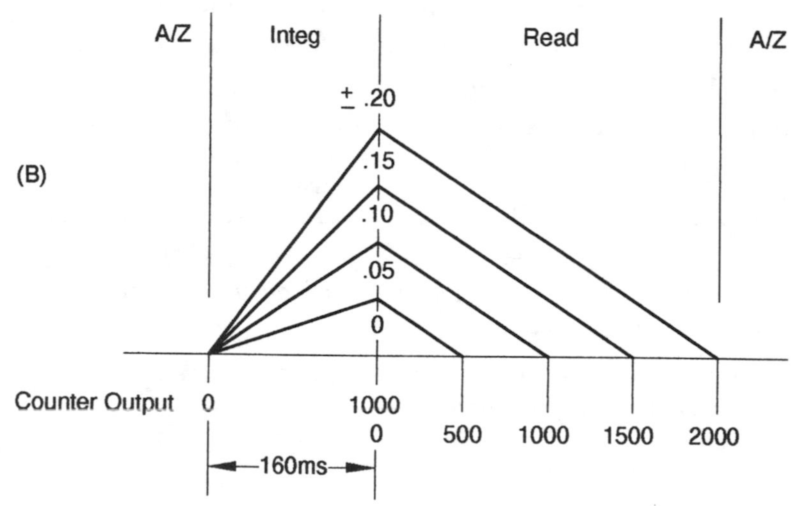

Figure 16-3B: Complete analog to digital conversion cycle.

The AZ period varies from 1000 to 3000 counts. The INTEG period is fixed at 1000 clock pulses (160 milliseconds) during which the unknown input voltage builds up to some level between 0 and 2000 mV. The greater the input voltage, the higher it will build during the fixed INTEG period.

The READ period is a variable time that is proportional to the integrated input voltage: the larger the input voltage, the longer the READ time. It can vary from zero counts for zero input voltage to 2000 counts for a full scale input voltage. The value of the voltage is then determined by simply counting the number of clock pulses that occurred during the READ period. For an input voltage less than full scale, AZ gets the unused portion of the READ period.

More Details

Look again at Figures 16-3A and 16-3B. During auto zero, a ground reference is applied as an input to the A/D converter. Under ideal conditions, the output of the comparator would also go to zero. However, input offset voltage errors accumulate in the amplifier loop and appear at the comparator output as an error voltage. This error is impressed across the AZ capacitor (C4, pin 29) where it is stored for the remainder of the measurement cycle. This stored level is used to provide offset voltage correction during the INTEG and READ periods.

The INTEG period begins at the end of the AZ period. As the period begins, the AZ switch opens and the INTEG switch closes. This applies the unknown input voltage to the input of the A/D converter. The voltage is buffered (R14, pin 28) and passed on to the integrator to determine the charge rate (slope) on the INTEG capacitor (C5, pin 27). At the end of the fixed INTEG period, the capacitor is charged to a level proportional to the unknown input voltage.

During the READ period, this voltage is translated to a digital indication by discharging the capacitor at a fixed rate and counting the number of clock pulses that occur before it returns to the original auto zero level.

As the READ period begins, the INTEG switch opens and the READ switch closes. This applies a known reference voltage to the input of the A/D converter. The polarity of this voltage is automatically selected to be opposite that of the unknown input voltage, thus causing the INTEG capacitor to discharge at a fixed rate (slope). This rate is determined by the known reference voltage (C2, pins 33 and 34). When the charge is equal to the initial starting point (auto zero level), the READ period is ended. Since the discharge slope is fixed during the READ period, the time required for discharge is proportional to the unknown input voltage.

The AZ period and thus a new measurement cycle begins at the end of the READ period. At the same time, the counter is released for operation by transferring its contents (the previous measurement value) to a series of latches. This stored data is then decoded and buffered before being used to drive the LCD display.

DC Voltage Measurement

Figure 16-4 shows a simplified diagram of the DC voltage measurement function. The input voltage divider resistors add up to 1 megohm. Each step down divides the voltage by a factor of ten. The divider output must be within the range –0.199 to +0.199 volts or the overload indicator will function. The overload indication consists of a 1 in the most significant digit and blanks in the remaining digits.

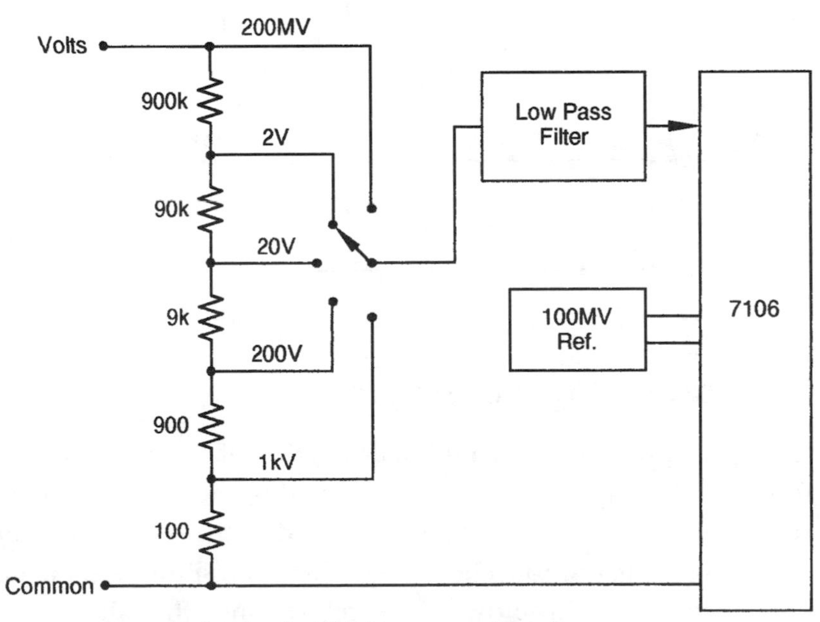

Figure 16-4: Using the DMM to measure DC voltage.

AC Voltage Measurement

Figure 16-5 shows a simplified diagram of the AC voltage measurement function. The AC voltage is first rectified and passed through a low pass filter to smooth out the waveform. A scaler reduces the voltage to the DC value required to give the correct root-mean-square (RMS) reading. The voltage is then fed to a 10 to 1 divider as in the DC voltage measurements.

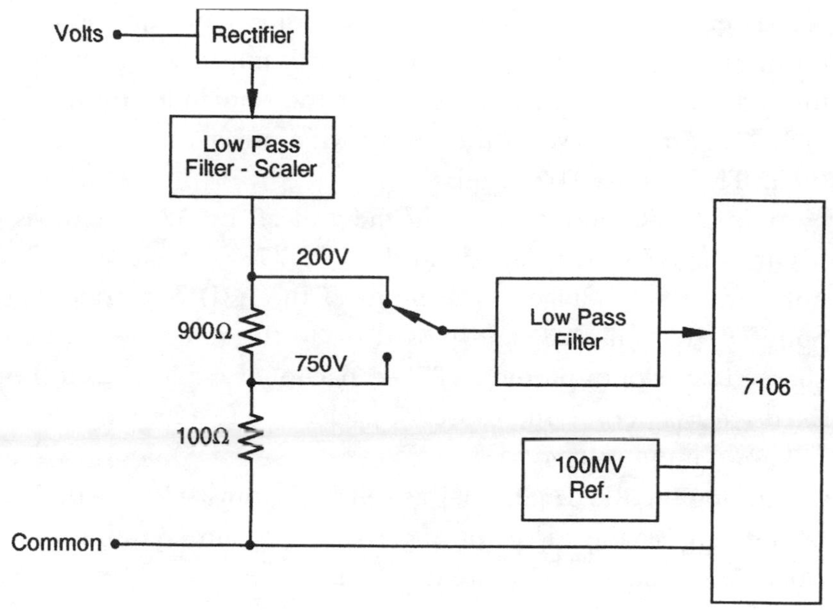

Figure 16-5: Using the DMM to measure AC voltage.

Current Measurement

Figure 16-6 shows a simplified diagram of the current measurement function. Internal shunt resistors convert the current to between –0.199 to +0.199 volts, which is then processed in the 7106 IC to light the appropriate LCD segments.

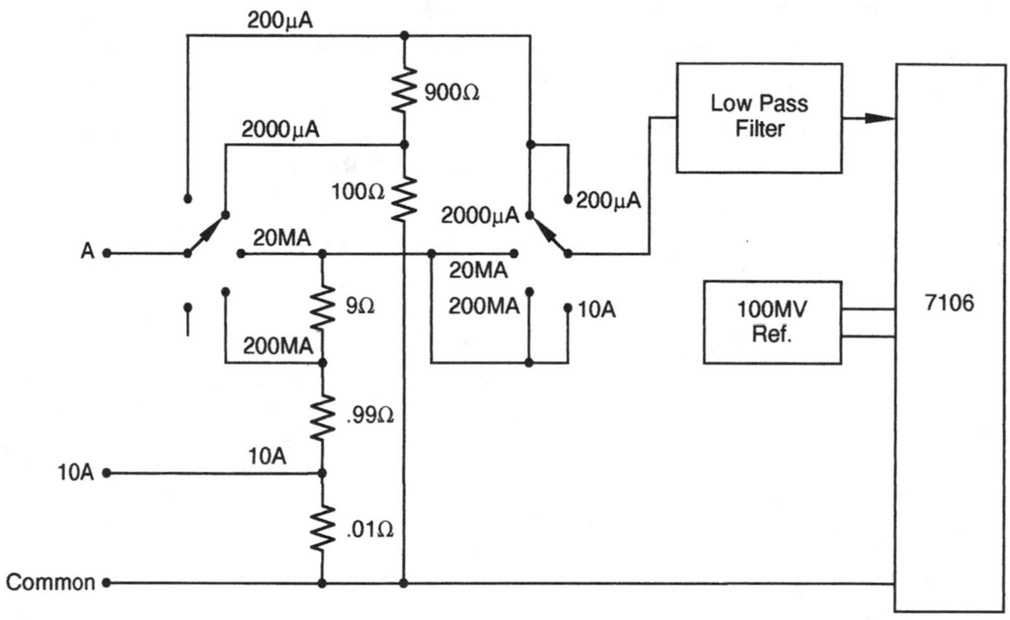

Figure 16-6: Using the DMM to measure current.

Resistance Measurement

Figure 16-7 shows a simplified diagram of the resistance measurement function. A simple series circuit is formed by the voltage source, a reference resistor from the voltage divider (selected by the selector switches), and the test (unknown) resistor. The ratio of the two resistors is equal to the ratio of their respective voltage drops. Therefore, since the value of one resistor is known, the value of the second can be determined by using the voltage drop across the known resistor as a reference. This determination is made directly by the A/D converter.

Overall operation of the A/D converter during a resistance measurement is basically as described earlier with one exception. The reference voltage present during a voltage measurement is replaced by the voltage drop across the reference resistor. This allows the voltage across the unknown resistor to be read during the READ period.

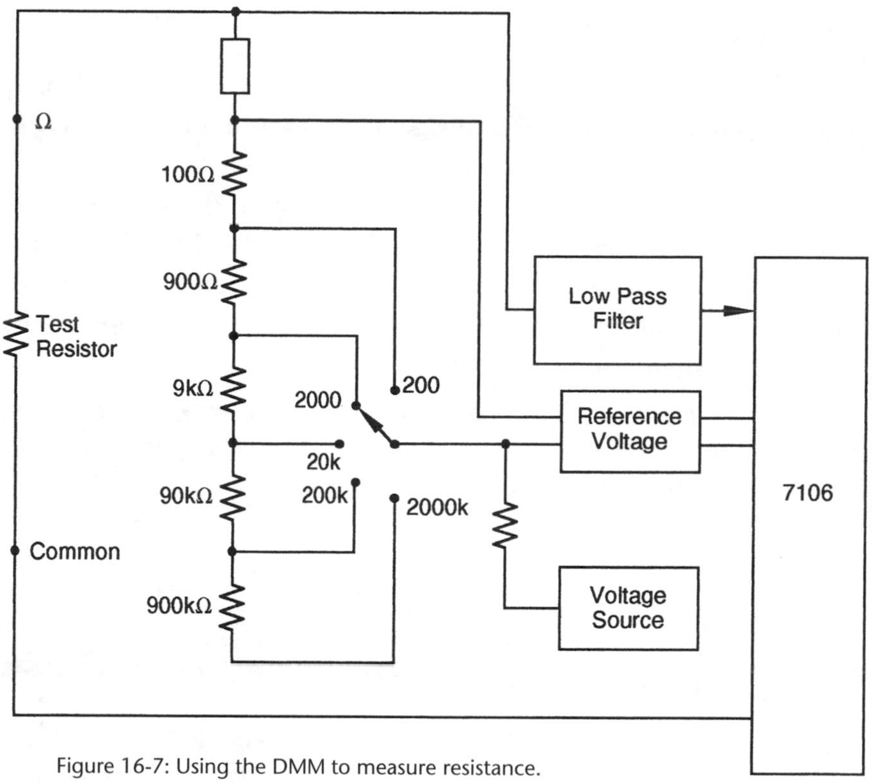

Figure 16-7: Using the DMM to measure resistance.

hFE Measurement

Figure 16-8 shows a simplified diagram of the common-emitter short circuit current gain (hFE) measurement function. Internal circuits in the 7106 IC maintain the COMMON line at about 2.8 volts below V+. When a PNP transistor is plugged into the transistor socket, emitter to base current flows through resistor R21. The voltage drop in resistor R22 due to the collector current is fed to the 7106 and indicates the hFE of the transistor. For an NPN transistor, the emitter current through R22 indicates the hFE of the transistor.

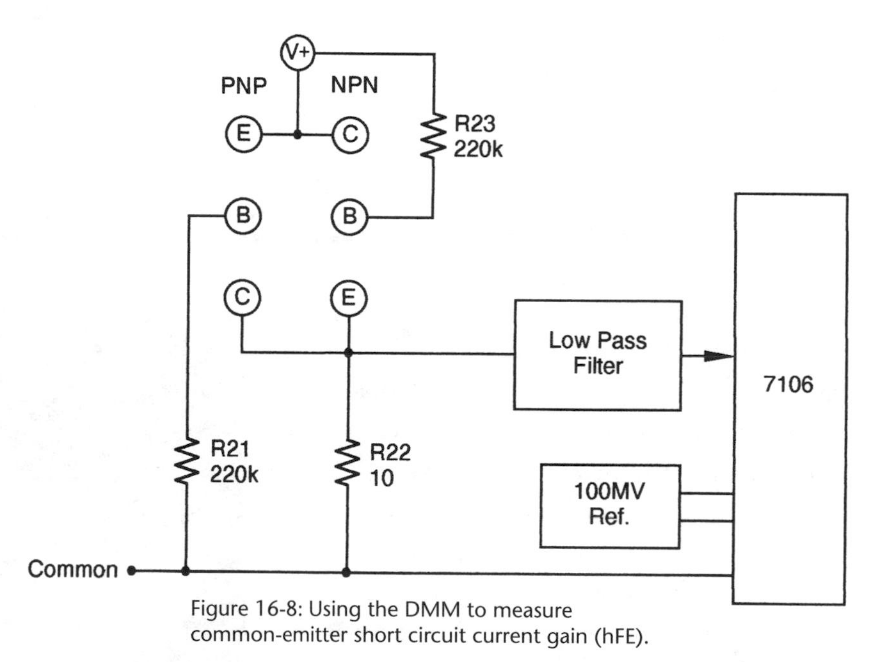

Figure 16-8: Using the DMM to measure common-emitter short circuit current gain (hFE).

Diode Check

Diodes are checked by applying a test voltage and measuring the forward voltage drop. With the positive DMM test lead connected to the anode of the diode, the meter should read between 450 and 800 mV. With the test leads reversed, a good or open diode will show overrange, while a shorted diode will show a low value, close to 000.

Construction

The instruction manual made the assembly of the M-1005K parts kit relatively painless. I say "relatively" since this was not a simple assembly. You must be very careful to get each of the components in the right place: incorrect resistor values will cause incorrect operation. Fortunately, the manual carefully gives the color codes for each resistor, and having each mounted individually on a printed cardboard strip reduces the chance of error. Similarly, the one diode and two transistors must be oriented properly, and the 40-pin integrated circuit must be properly oriented and inserted into its socket.

Many of the parts are very close together, and it is easy when soldering to bridge two leads that should not be bridged! I did this in two places, which gave me grief later. Unfortunately, the manual does not show the two-sided printed circuit board layout, so it is sometimes difficult to determine where leads can be soldered together without a bridging problem. Figure 16-9 shows the interior of the unit; the pencil points to the 7106 IC.

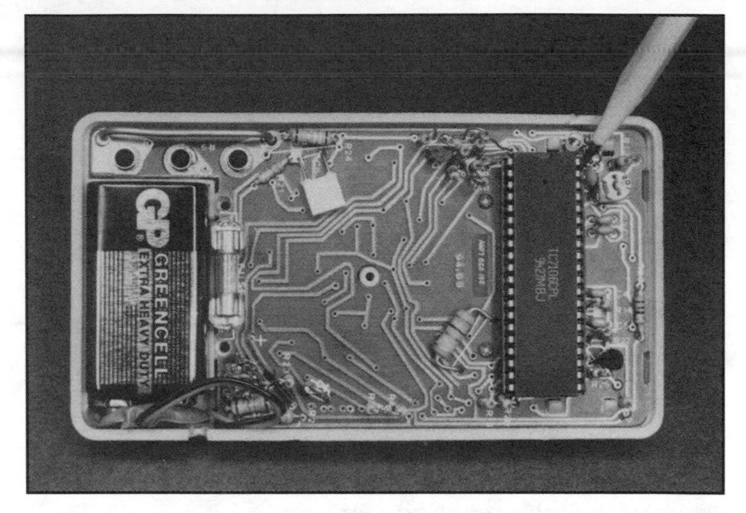

Figure 16-9: Interior view of the DMM; the pencil points to the 7106 IC.

After all the soldering is done, you assemble the display to the printed circuit board. A special plastic frame with four legs is used to sandwich the LCD window plate and two contact strips (called "zebras") to the printed circuit board. The stiff legs of the frame can easily be bent and broken, so do this carefully.

The trickiest part of the assembly is the mounting of the selector knob. First you insert six small formed metal slide contacts to the underside of the knob, then use the supplied grease and place two small springs in two holes on the top of the knob. Then you place two small steel balls into opposite detents in the case cover, invert the knob, and place the knob springs onto the balls. When the printed circuit board is placed over the back of the knob and fastened down, the knob should turn smoothly and stop positively in each detent position.

I had trouble with this. When I turned the knob, one of the steel balls would pop out. This happened several times. I had to use more grease and stretch the ball springs a bit before the knob turned smoothly and stopped positively at each detent.

The result of all this effort is a very professional-looking unit. Figure 16-10 shows the final assembled DMM.

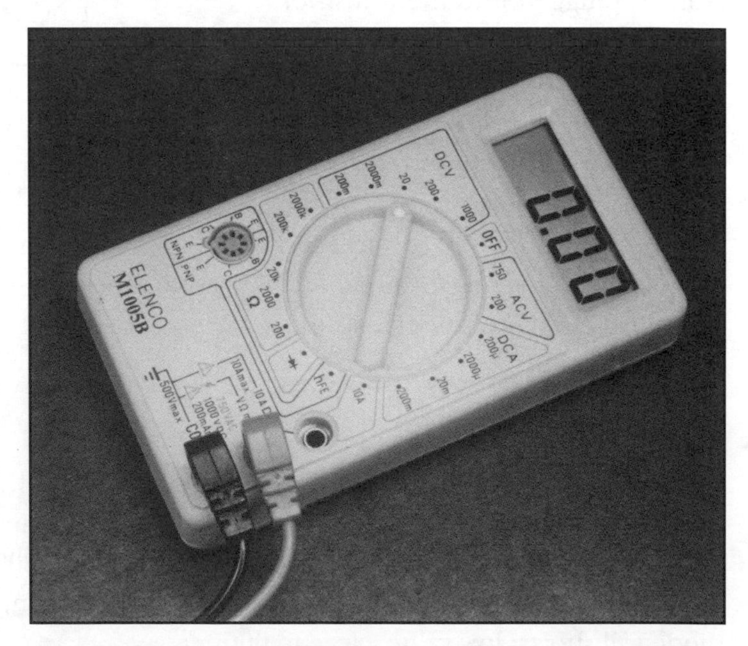

Figure 16-10: The completed DMM unit.

Testing

Finally, the moment of truth! I snapped the 9 V battery terminals to the supplied battery, and switched the selector knob from OFF to the first position clockwise position, 750 V AC. Instead of the display showing 000, it showed 015. As I switched clockwise around to the hFE position, the only change was that the decimal point was either not there at all, or in one of two locations.

At the diode test position, a 1 appeared on the display by itself. In the 200 ohm switch position, a 1 appeared with a decimal point before the final blank digit. Around the rest of the switch positions there were various displays, but none showing 000, which is what it should show in all positions except off!

Hmmmm. This was bad. The unit did not auto-zero. I tried adjusting the one potentiometer, VR1, but it had no effect.

The manual makes no mention of this, but there is an easy way to test the LCD display. With a jumper between pin 37, the TEST terminal of the 40-pin 7106 IC, and pin 21, the LCD backplane terminal, the display should light ALL the LCD segments except the decimal points, showing –1888 on the display whenever the selector switch is in any position except OFF. I did this, and I got the –1888 display, showing that at least all the display drivers in the 7106 IC were operating.

Troubleshooting and Scopes

I performed a very careful inspection with a magnifying glass, checking the value of all components, and the soldering. I found no parts out of place, but there were two places where solder had bridged where it shouldn't have. I confirmed this by referring to the schematic and using an ohmmeter to show continuity where there should be none! I cleared the solder bridges, and convinced this was the problem, I tested the DMM again—with the same bad results!

It was time to use my oscilloscope to see if I had the proper trace on each pin. Although the manual did not show these traces, I had a properly working DMM using the same 7106 IC. Its manual showed the proper traces on some significant pins, so I compared traces. I found that six pins (27, 28, 29, 33, 34, 35) on the defective unit were drastically different from the properly working unit. This certainly made the 7106 IC in the new unit suspect.

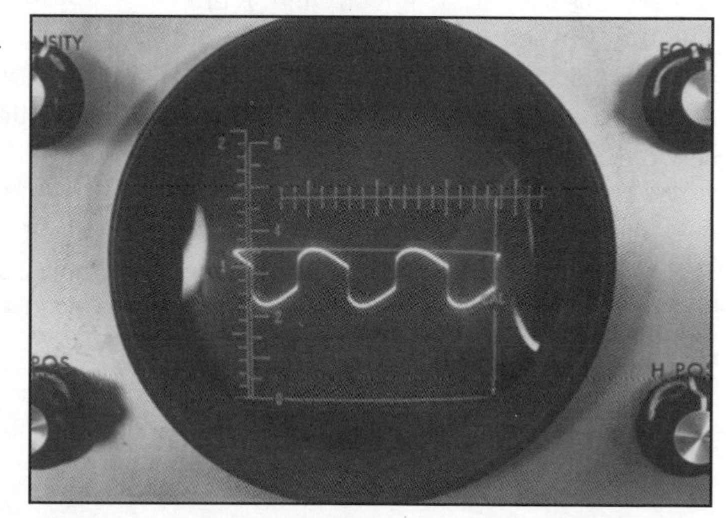

Figure 16-11: Distorted pin 40 signal displayed on the old 5 MHz scope.

I should note here that I used two analog scopes during this troubleshooting. The first, an old RCA WO-33A with a low 5 MHz bandwidth (adequate 40 years ago) showed traces that did not look like they should. Figure 16-11 shows this display. However, when I used a modern

B+K Model 2120B 25 MHz scope, the traces were just like the manual that showed the traces. See Figure 16-12. The lesson here is that digital signals, except at low frequencies, do not display properly on low bandwidth scopes!

I carefully checked the circuitry feeding the 7106 and could find no errors, so I called C&S Sales, the kit source, and told them my problem. They sent me a replacement 7106. When I used this to replace the original 7106, the meter worked perfectly!! Was the original 7106 defective, or did my two solder bridges destroy some internal functions? We'll never know. . .

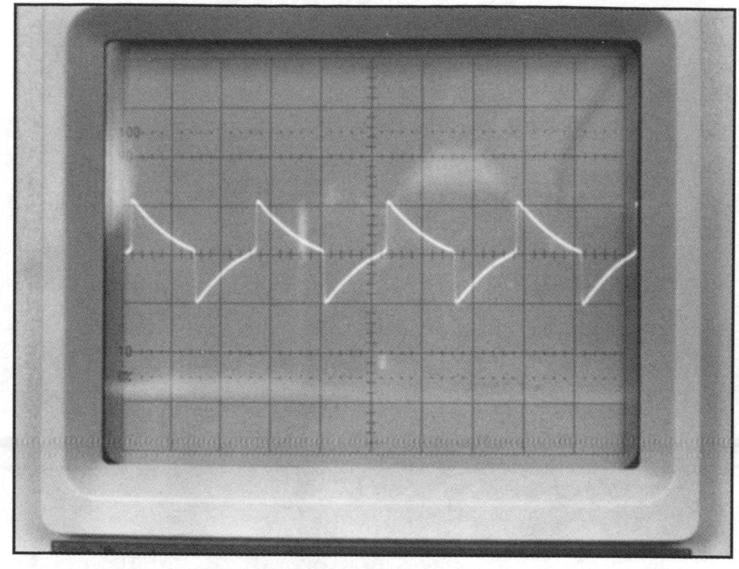

Figure 16-12: The correct "look" of the pin 40 signal as seen on a 25 MHz scope.

Now it was simply a matter of calibrating the DMM by feeding in a known voltage around 10 volts DC (using a known accurate digital voltmeter for reference) to the DMM with the switch in the 20 V DC position, and adjusting VR1 so that the display shows the known voltage. All other ranges automatically adjust with this calibration.

The M-1005K uses very little battery power, and none when it is OFF. With the switch in the diode test or 200 ohm positions, about 3.5 milliamperes is drawn from the 9 volt battery. In all other operating positions, only from 1.3 to 1.5 milliamperes are used. This should give the battery many tens of hours of use.

Final Comments

After assembling the M-1005K and getting it working properly, I tested resistors, diodes, transistors, power supplies, etc., using the supplied leads. The unit worked well, and readings compared very favorably with my other more expensive DMMs. If your needs are not critical, your budget is modest, and you like building and troubleshooting electronic kits, you should be very pleased with this project.

The M-1005K digital multimeter parts kit is available from C&S Sales, 150 W. Carpenter Ave., Wheeling, Illinois, 60090. The price at the time this was written was $19.95 plus $5 shipping. Call 800-292-7711 to order. Pricing and availability may have changed since publication of this book.

CHAPTER 17

A Transistor/Diode Tester

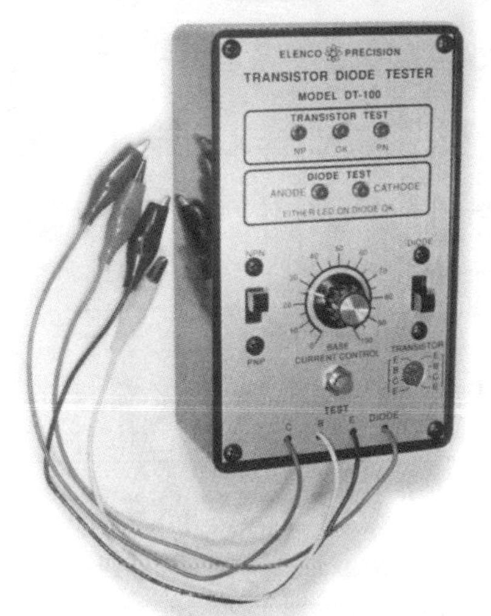

The best way to test transistors and diodes is through "dynamic testing." This chapter describes the theory and use of a dynamic transistor and diode tester. Since many of the necessary parts are hard to find (and expensive!) individually, this chapter will describe construction using a parts kit.

Transistors and diodes are used in just about every electronic circuit these days, even those that use integrated circuits as their main elements. Diodes are typically used to direct current flow while transistors are often used as amplifiers and drivers. While simple static "go/no-go" testing can be performed with an ohmmeter and some knowledge of function, dynamic testing actually puts the component to use. The transistor/diode tester described in this chapter uses dynamic testing. An 8-pin socket allows testing of smaller transistors with any of the six possible lead positions; three colored external clip leads are used for larger components. A fourth clip lead is used for diode testing. A pushbutton TEST switch is operated when the component is connected and the NPN/PNP and DIODE/TRANSISTOR slide switches are set. The results are displayed by the proper glowing of three LEDs for the transistor test, or two LEDs for the diode test. A rotary BASE CURRENT CONTROL provides a means for roughly matching the gain of similar transistors. Power is provided internally by a common 9 volt battery.

The DT-100 can check most types of diodes—germanium, silicon, power, LED and Zener—out of the circuit, and in-circuit with resistors as low as 5 kilohms. The diode test also automatically identifies the anode and cathode. Most types of transistors (germanium, silicon, power, RF, audio, switching, FET, etc.) can be tested out of the circuit, and in-circuit with base or collector resistors as low as 100 ohms. The transistor test also identifies NPN and PNP types, and the relative gain of two transistors.

Building with the Parts Kit

This project is based upon the Elenco DT-100K parts kit available from C&S Sales (ordering information is given at the end of this chapter). The parts kit includes a printed circuit board and all components. An instruction manual for building the tester is included, so we will not go into construction details here.

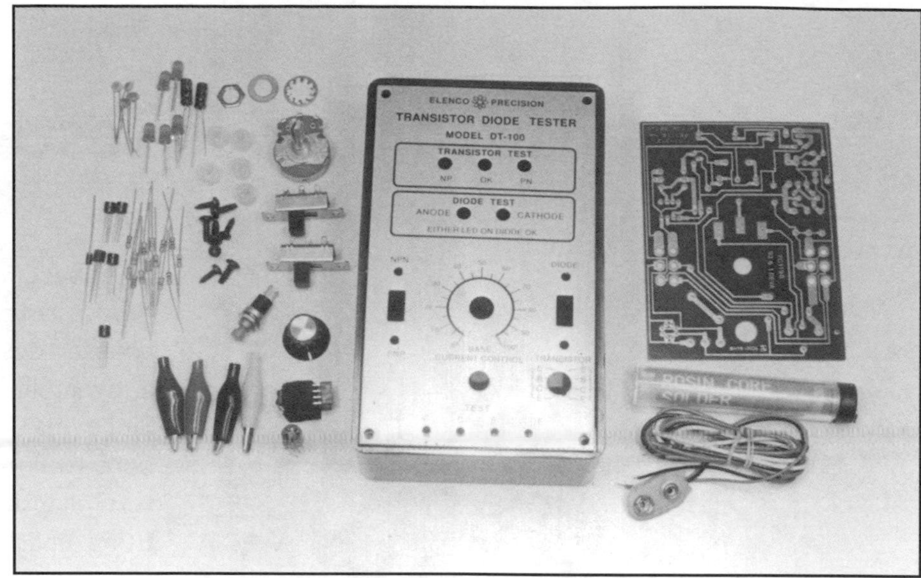

Figure 17-1: The parts kit for building the transistor diode tester.

I found only two minor problems assembling the kit. The instructions did not call for mounting the printed circuit board subassembly to the front panel until the top of Page 6; it makes more sense to do so at the top of Page 5. Also, when I tried to mount the sub-assembly to the front panel using the four #4 screws provided, they were too small and just slipped through the mounting holes. A quick phone call to Elenco's responsive parts department (provided on a card included with the kit) solved the problem with four #4-40 screws and nuts.

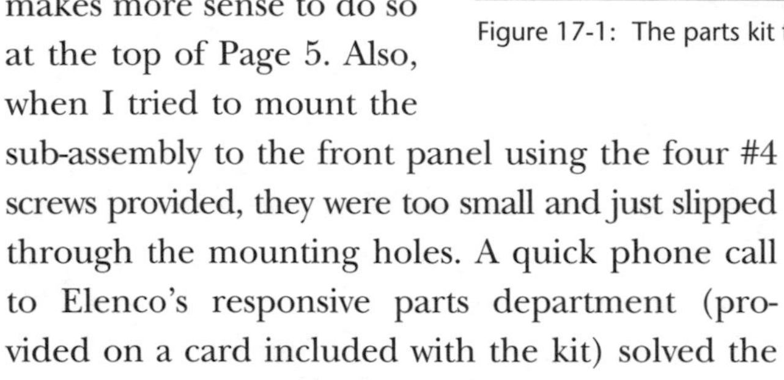

Figure 17-1 shows the contents of the parts kit. Notice the attractive housing that is included. Figure 17-2 shows the assembled unit.

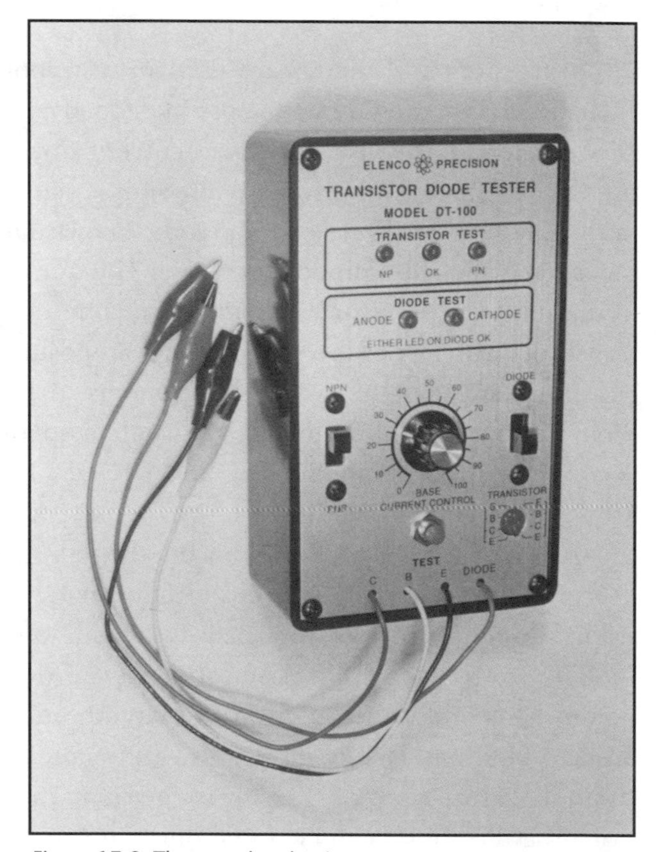

Figure 17-2: The completed unit.

Circuit Description

Figure 17-3 shows the schematic of the transistor/diode tester. SW1 is the NPN/PNP slide switch, shown in the NPN position. SW2 is the DIODE/TRANSISTOR slide switch, shown in the TRANSISTOR position.

For the purpose of explaining circuit operation, assume an NPN transistor is plugged into the TRANSISTOR TEST socket (or using the external C, B, and E test leads), and that the leads are properly oriented—that is, that the test transistor collector, base, and emitter leads are properly connected to the C, B, and E test points respectively. Switch SW1 is in the NPN position, and SW2 in the TRANSISTOR position.

When the pushbutton TEST switch is closed, positive voltage is applied through diode D1, through SW2, and through SW1 to the top of BASE

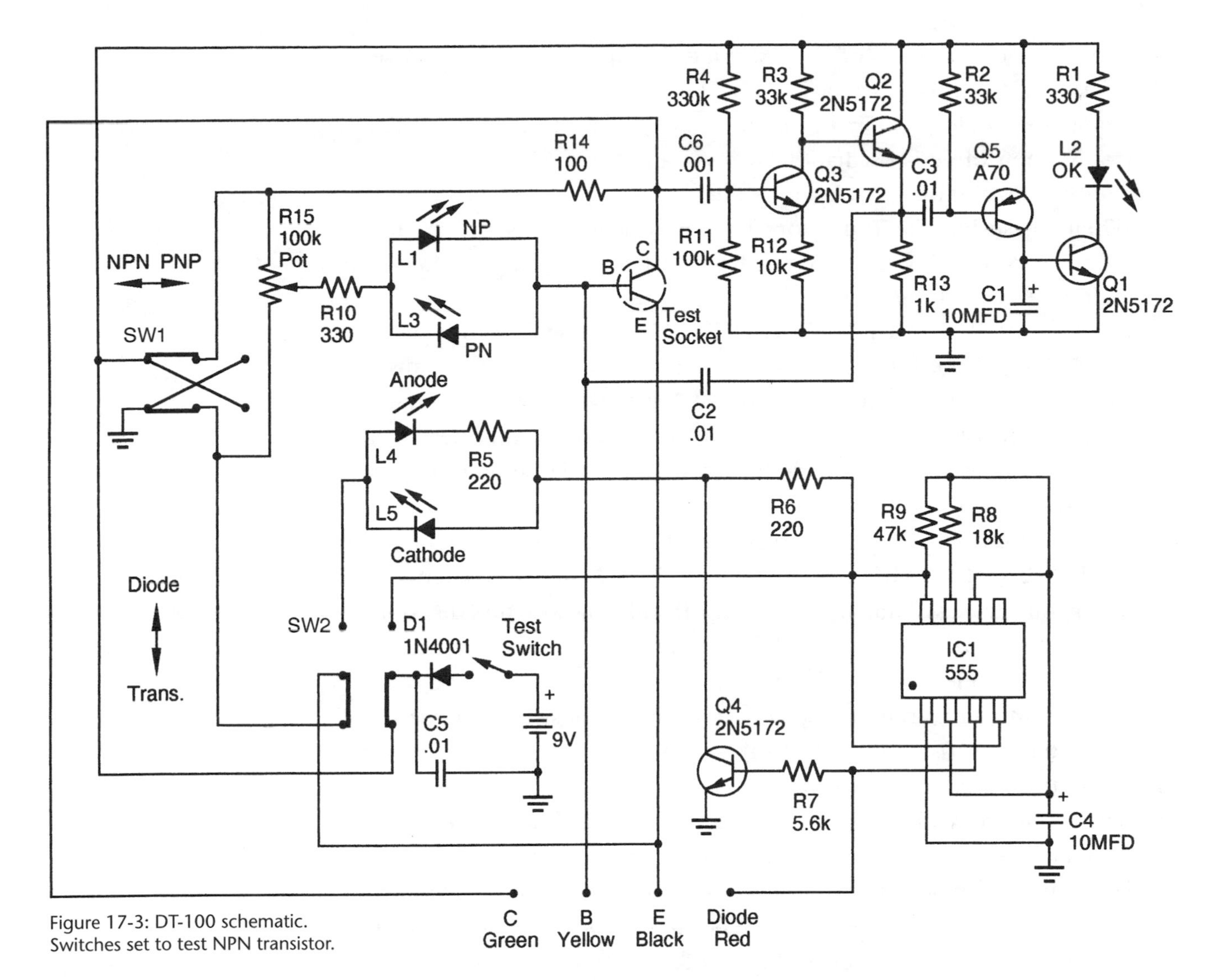

Figure 17-3: DT-100 schematic.
Switches set to test NPN transistor.

CURRENT CONTROL potentiometer R15, and (through resistor R14) to the collector of the test NPN transistor. This also positively charges capacitor C6. Positive voltage also goes to resistors R1, R2, R3 and R4, as well as the collector of NPN transistor Q2, and the emitter of PNP transistor Q5.

If the setting of R15 allows enough test transistor base current to flow through resistor R10 and NP LED L1 (even though L1 may not glow perceptively), collector-to-emitter current will flow. Since the emitter is connected to circuit ground through SW2 and SW1, the voltage at the collector drops toward ground, quickly discharging capacitor C6, after which it recharges more slowly through R14.

The dynamic testing is accomplished by making the test transistor operate in an oscillator circuit. NPN transistors Q2 and Q3, and their associated resistors (R3, R4, R11, R12, R13), are wired as amplifiers, triggered by the fast discharging and slower charging of C6. The output of Q2 at its emitter is fed back to the base of the test transistor through capacitor C2. This positive feedback maintains oscillation. The frequency of oscillation increases up to about 20 kHz as base current to the test transistor is increased.

At the same time, part of the Q2 emitter output is fed to the base of PNP transistor Q5 through capacitor C3. Here resistors R2, R1, capacitor C1 and transistor Q1 form a driver circuit to switch on the OK LED (L2) to light only when there is oscillation. If test transistor saturation is reached, oscillation stops and the OK LED goes out.

When measuring PNP transistors, the power polarity supplied to the test transistor is reversed via the NPN/PNP switch, so base current flows through PN LED (L3) instead of L1.

Rotating the BASE CURRENT CONTROL clockwise from full counterclockwise changes the voltage at the wiper, and increases the test transistor base current. The lower the base current for the OK LED to glow, the higher the gain (beta) of the transistor under test. Comparative tests of the gain of two transistors can be made by observing the dial setting to provide similar intensity of the glowing OK LED.

It should be noted that as the base current is increased by rotating the BASE CURRENT CONTROL clockwise, the PN LED will glow when a PNP transistor is under test. Similarly, the NP LED will glow with an NPN transistor under test. In both cases, the OK LED may go out if the test transistor enters saturation and stops oscillating.

The design configuration is such that in-circuit transistors can be tested provided that the base and collector resistors are greater than 100 ohms. However, it can often be difficult to access the leads of in-circuit transistors, and even more difficult to determine sneak circuit paths to other components that may lower the base or collector resistance below 100 ohms.

For testing diodes, slide switch SW2 is set in the DIODE position. SW1 may be in either the NPN or the PNP position. When the TEST switch is pushed, positive voltage is applied through diode D1 and SW2 to power pin 8 of IC1, a 555 timer/oscillator integrated circuit. Resistors R8 and R9, and capacitor C4, set the 555 to alternate from about supply voltage (high) to circuit ground (low) at output pin 3 about twice per second.

To explain circuit action, assume a test diode is connected so that the red DIODE test lead is clipped to the test diode anode, and the E test lead is clipped to the cathode of the test diode. As IC1 oscillates, every time a HI appears on pin 3 of IC1 the positive voltage travels through the test diode and appears (through SW2) at ANODE LED L4. Positive voltage also appears at the base of NPN Q4 through resistor R7, allowing Q4 to conduct and effectively bringing its collector to ground. This allows L4 to light through dropping resistor R5. In the next half-cycle of IC1, when pin 3 drops to ground, Q4 is cut off, and no positive voltage at the test diode anode means it is not conducting. L4 goes out, then comes on again when pin 3 of IC1 goes HI in the next half-cycle. In other words, ANODE LED L4 blinks on and off.

If the test diode is connected so the DIODE test lead is clipped to the CATHODE and the E test lead is clipped to the ANODE, the circumstances are changed. When the test switch is closed, positive voltage from the battery, through diode D1 and switch SW2, appears at dropping resistor R6 and CATHODE LED L5. Since L5 is connected to the E test lead through SW2, whenever the output pin 3 of IC1 goes LOW, it provides a ground through the test diode to light L5. When pin 3 goes HI, transistor Q4 conducts, effectively removing positive voltage at L5, so it goes out. Again, the result is a blinking LED, but this time it is the CATHODE LED blinking, which means the red test DIODE test lead is connected to the cathode of the test diode. Simple, but ingenious!

By just connecting a diode between the DIODE and E test leads, the blinking LED will tell you if it is working properly, and its polarity.

All types of diodes may be tested: silicon, germanium, LEDs, or Zeners over 6 volts. Zener diodes under 6V cause the second LED to glow at lower intensity, indicating that Zener breakdown has occurred.

Testing Your Assembled Unit

Once you've assembled the transistor/diode tester from the DT-100 parts kit, you should perform the following tests to be sure it's operating properly.

A standard 9 V 2U6-type radio battery is used inside the tester. You simply connect it to the battery clip and anchor the battery to the bottom of the case with double-sided tape. The battery should be good for many hours of testing.

DIODE Circuit Tests: Place the DIODE/TRANSISTOR switch in the DIODE (up) position. Connect the red and black clip leads together and push in the TEST button. The diode test LEDs should blink on and off at about a 1 Hz rate. Next, connect the red and black leads to any good diode. Only one LED should blink. Reversing the leads should cause only the other LED to flash.

TRANSISTOR Circuit Tests: Place the DIODE/TRANSISTOR switch in the TRANSISTOR (down) position. Short the yellow (B = base) and black (E = emitter) leads together and rotate the BASE CURRENT CONTROL fully clockwise. Press the TEST button. With the NPN/PNP switch set to the NPN position, the NP LED should light. With the switch set to the PNP position, the PN LED should light.

Rotate the BASE CURRENT CONTROL to the fully counter-clockwise position. Set the NPN/PNP switch to NPN, keeping the DIODE/TRANSISTOR switch in the TRANSISTOR position. Next place a known good NPN transistor in the test socket with the emitter in E, base in B and collector in C, or use the external clip leads. Be sure none of the leads are shorting. Press the TEST switch and slowly rotate the BASE CURRENT CONTROL. The OK LED should light, indicating the test transistor is oscillating. The higher the gain of the transistor, the less you should have to rotate the control before the OK LED comes on. As you rotate the control further clockwise, the NP LED will come on as the base current increases, and the OK LED might go off.

Repeat this test with the NPN/PNP switch in the PNP position, and using a known good PNP transistor. The results should be the same, except the PN LED will come on as the control is turned clockwise.

Using the Transistor/Diode Tester

Here are some step-by-step procedures for using your transistor/diode tester.

To test diodes:

1. Place the DIODE/TRANSISTOR switch in the DIODE position.
2. Connect test diode to red and black leads.
3. Press TEST button. One diode LED (ANODE or CATHODE) should blink. The blinking LED tells you which lead of the test diode is connected to the red DIODE test lead.
4. If both LED lamps blink, then the test diode is shorted.
5. If neither lights, then the test diode is open.

To test transistors out-of-circuit:

1. Place DIODE/TRANSISTOR switch in TRANSISTOR position.
2. Place the NPN/PNP position in the proper position for the transistor to be tested. If not sure, see later step.
3. Place test transistor in socket or attach to C, B, and E leads. If collector C, base B, and emitter E are not known, assume B is the center lead on small plastic transistors and C the metal case or tab on power transistors.
4. Rotate the BASE CURRENT CONTROL fully counter-clockwise.
5. Press TEST button. If either the NP or PN LED come on, you probably have the NPN/PNP switch in the wrong position! Change it. (If the OK LED glows, this indicates a good transistor.)
6. You will probably need to rotate the BASE CURRENT CONTROL clockwise so that the OK LED glows. This indicates a good transistor. Continue to rotate the control and the proper NP or PN light should come on. The OK LED may go out.
7. If no LEDs glow, the transistor is bad—or, more probably, you have not identified the leads properly. Repeat assuming other lead arrangements.
8. When the transistor is shown to be OK, the base current control gives an indication of transistor beta. The lower the setting relative to another transistor, the higher the beta.

Transistor Testing (In Circuit)

The transistor/diode will test transistors in circuit provided the base biasing resistance is greater than 100 ohms. Simply follow the previous procedure for testing out of circuit transistors. Do not apply power to the circuit of the transistor under test because the tester supplies the necessary power.

If you know the "gender" and lead arrangement of the transistor under test, the transistor/diode tester confirms operation, and even allows you to match transistors with nearly the same gain. If you are not sure whether you have an NPN or PNP under test, and when you are not sure which leads are the base, collector, or emitter, things get confusing. Just follow the test procedure(s) described above. When things are set properly, the LEDs will tell you so.

The Elenco DT-100 transistor/diode tester parts kit is available from C&S Sales, 150 W. Carpenter Ave., Wheeling, IL 60090. The price at the time this was written was $22.95. Add $5 for shipping, and sales tax if shipping to Illinois. Call 800-292-7711 to order. Pricing and availability may have changed since this book was published.

CHAPTER 18

Quick Tracey, The Semiconductor Sleuth

When used with any oscilloscope that has both vertical and horizontal inputs, this simple circuit allows you to show distinctive traces for most semiconductors as well as many other types of electronic parts. No special circuit board is needed.

In the "lineup" were Danny Diode, Zachary Zener, Sammy Selenium, "Silly" Con Rectifier, Tommy Transistor, Usiah Unijunction, and Phineas Photoconductor. The problem: which one was the "bad guy"? Clearly a case for that surreptitious semiconductor sleuth, "Quick Tracey"!

Using your oscilloscope and Quick Tracey, it's a simple, quick job to wade through various types of "bargain" semiconductors and sort out the bad ones by interpreting a trace on your scope screen. Quick Tracey's regular "report" can include checking transistors for polarity (PNP or NPN), approximate gain, and linearity. Also, all tests are dynamic rather than static, and will thus uncover some defects that static tests won't show.

When testing diodes, shorts or opens show up like the proverbial sore thumb; so does reverse polarity. On low-voltage Zeners, not only will you be able to tell if they are good or bad, but you'll also be able to estimate their breakdown voltage. Even the less common semiconductors, such as unijunction transistors, tunnel diodes, and silicon-controlled-rectifiers, are cases easily handled by Quick Tracey. It costs about $20 to build from all new parts (and less if you have some common parts in your junk box), and even has its own built-in calibration circuit.

Circuit Description

Figure 18-1 is the schematic of the complete Quick Tracey. Since this actually contains three circuits (calibrate, diode test, and transistor test) I have shown each circuit separately in Figures 18-2, 18-3, and 18-4 to simplify the explanation of circuit operation. Although Quick Tracey can be used without any understanding of the circuit theory, you'll find other uses for the unit if you are familiar with its "modus operandi" (that's Latin for "how da' t'ing woiks").

Figure 18-2 shows the power and calibration circuit. When normally-open pushbutton switch S1 is depressed, equal voltages appear across Rl and R2 (since they have the same resistance value), thus giving equal deflection voltages across the scope vertical and horizontal inputs. (The scope sweep selector must be set to "horizontal input" or "external sweep"). By proper adjustments of the scope vertical and horizontal gain controls, you will get a sloping 45 degree line on the screen. This, in effect, sets the scope controls for equal gain on the vertical and horizontal inputs.

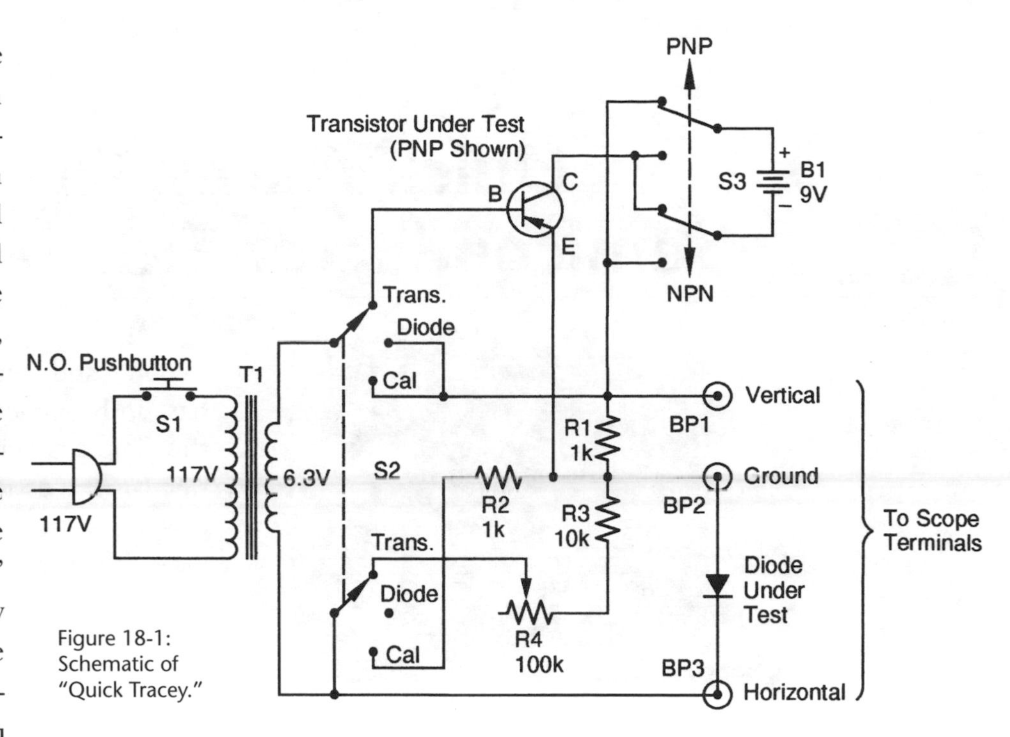

Figure 18-1: Schematic of "Quick Tracey."

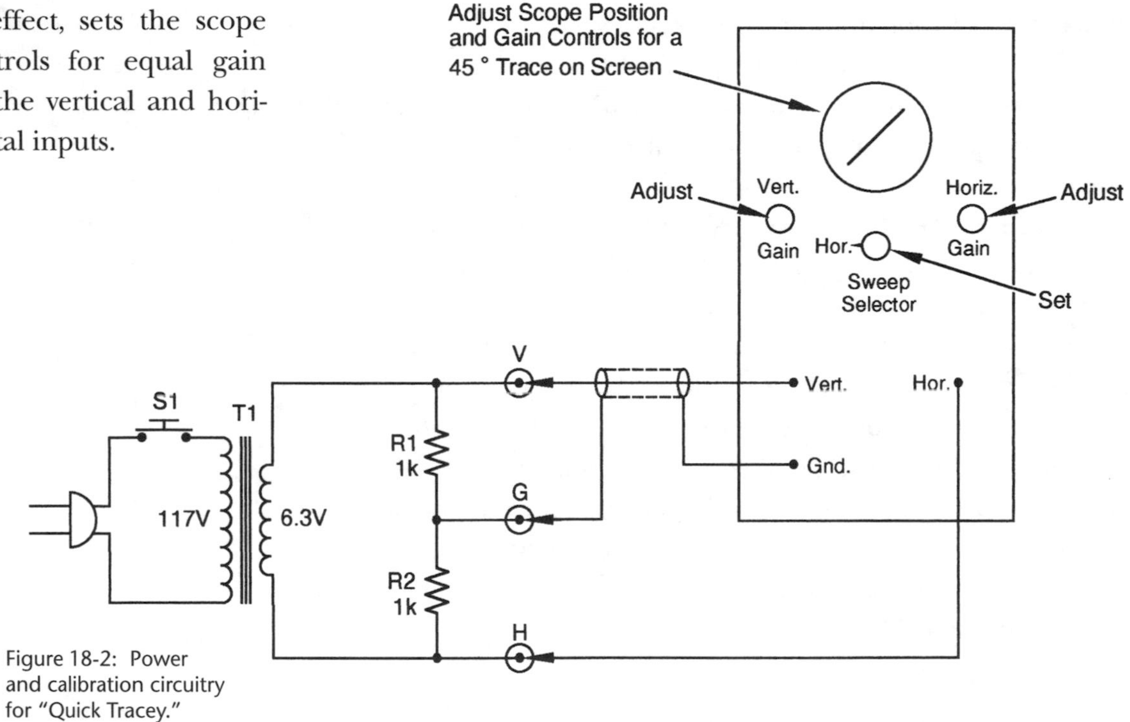

Figure 18-2: Power and calibration circuitry for "Quick Tracey."

IMPORTANT NOTE: The 45-degree slant shown in the Figure 18-2 illustration, leaning to the right, may lean to the LEFT on your scope. If this is so, then ALL your scope traces will be the opposite (mirror-image) of the illustrations in this chapter. This has to do with your scope's horizontal deflection (left-to-right or right-to-left), which varies with different manufacturers.

Figure 18-3 shows the diode test circuit. Think of the diode under test as a switch; when it's conducting (forward-biased) it's like a closed switch, and when it's reverse-biased it's like an open switch. Now when we apply 6.3 volts AC, we are alternately opening and closing this "switch" (the positive half-cycle forward biases the diode, the negative half-cycle reverse biases the diode). When the diode is conducting, it's the same as if we had shorted the horizontal scope terminal to the ground terminal, and the full voltage appears across Rl. The scope shows only a vertical line under this condition. However, on the other hand, when the diode is not conducting, there is NO current flowing through Rl, therefore NO vertical deflection, but FULL horizontal deflection. (The scope, remember, draws only infinitesimal current at 60 cycles, 6.3 volts). When the recurrent half-cycles are combined in the scopc trace, the pattern is half vertical and half horizontal for a perfect diode. The poorer the diode, the less perfect the pattern. Representative traces for typical cases are shown in Figure 18-3.

When testing a low-voltage Zener diode, the horizontal leg will "break down" at some distance out from the junction if the Zener is rated at less than 10 volts. Higher voltage Zeners can only be checked on Quick Tracey for diode action, but not Zener effect.

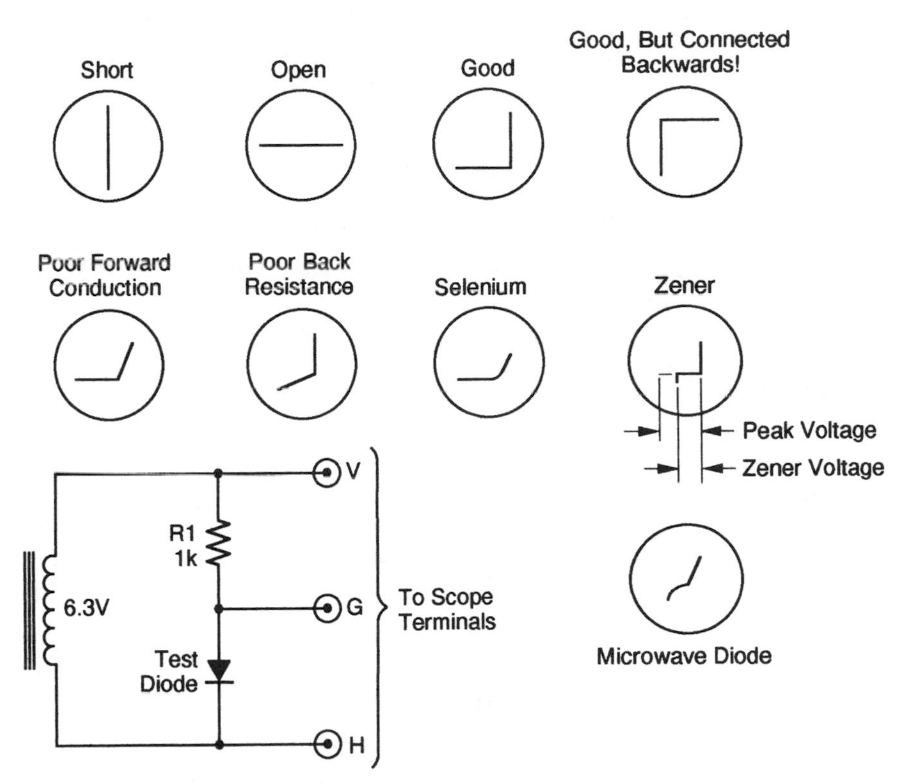

Figure 18-3: Diode test circuitry for "Quick Tracey."

Poor diode back resistance shows up on the trace as a down-ward slanting of the horizontal leg; with poor forward conduction, the vertical leg slants to the right. Selenium rectifiers, for example, usually show a relatively high forward resistance, high voltage drop (short vertical leg) and poor recovery characteristics (rounded junction of horizontal and vertical trace). All testing done with the Quick Tracey is done at a very low power level; there is no danger of harming the unit under test. Even the touchy 1N23 microwave crystal diodes are undisturbed by Quick Tracey's investigation, although they do yield a peculiar trace (see Figure 18-3 microwave diode pattern). The 1N23 is a low-voltage low-current diode, and the lower curved leg shows breakdown (though controlled, therefore not damaging) at the test voltage. Later on I'll show you how to use the diode test circuit for other tests.

Figure 18-4 shows the PNP transistor test circuit. With a PNP transistor under test the emitter has positive battery voltage applied through Rl, and the collector is at negative battery potential. However, unless there is current flow in the base-emitter circuit, only a very small leakage current flows in the collector-emitter circuit; that's what transistors are all about. Notice that the base is directly connected to one side of the 6.3 volt AC supply, and the emitter is connected to the other side through R3 and R4. Therefore, whenever the alternate half-cycles make the emitter positive with respect to the base, emitter-base current flows through R3 and R4 (R3 is used for current-limiting when R4 is set at zero). This current flow is measured as a voltage across R3 and R4 at the horizontal scope terminals, and is a measure of the transistor INPUT current. Since we are applying AC this voltage is constantly varying. Now, since the collector-emitter circuit is forward biased by B1, when base current flows it follows that collector current will flow simultaneously through Rl. This is the OUTPUT current, which is read as a voltage at the vertical scope terminals. This is exactly synchronous (in step) with the input current, which CONTROLS the output current.

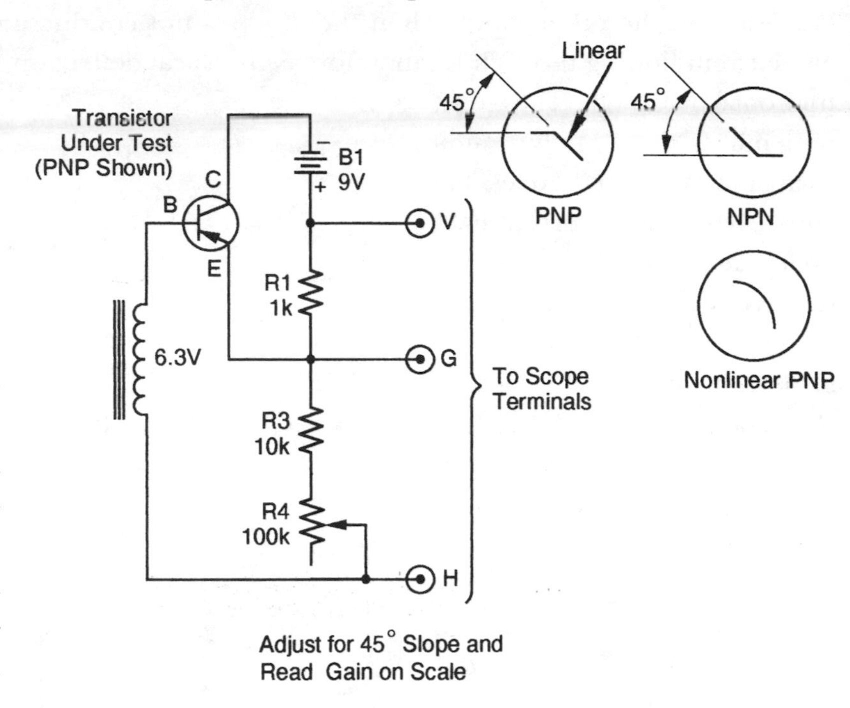

Figure 18-4: PNP transistor test circuitry of "Quick Tracey."

What does all this mean? Well, remember we calibrated the scope for equal vertical and horizontal deflection back in Figure 18-2, and now we use this fact to set our transistor-test scope trace slope to 45 degrees, using the R4 gain control. When the slope is 45 degrees, it means that the "input" and "output" VOLTAGES are equal. However, the voltages are dependent on the current flow through resistors Rl, R3, and R4. Remember Ohm's Law? If R4 is set at zero to get a 45 degree slope, then there is ten times the current flowing through 1K output resistor Rl than flowing through 10K input resistor R3 to make their voltage drops equal. Plainly and simply, the output current is ten times the input current, so the transistor has a "beta" (current gain) of 10. As the value of R4 is increased to set the trace slope to 45 degrees the ratio of output current to input current goes up. In other words, the transistor gain is higher. Using a numbered dial plate under the R4 control knob, you can read the approximate gain directly.

For NPN transistors, the theory of operation is identical, except that all polarities are reversed. This results in a reversed (inverted) scope pattern as compared with a PNP trace. This allows easy identification of an unknown, unmarked transistor.

Construction

Figure 18-5 shows my original unit, which was built over 30 years ago and still functions perfectly! None of the wiring is critical, so you can decide which features you want to incorporate. You may have many of the parts in your junk box.

First of all, decide on the cabinet you will use. If you use an aluminum cabinet (the Radio Shack #270-238 should do nicely if you use a small transformer for T1) be certain that your 117 VAC input voltage is isolated from the cabinet. Placing the gain control, polarity switch, binding posts, and function switch on the front panel is certainly logical. The power switch (S1) could be a toggle or slide switch instead of the pushbutton specified in the parts list; we preferred a pushbutton to insure that the unit was OFF except when actually viewing a trace.

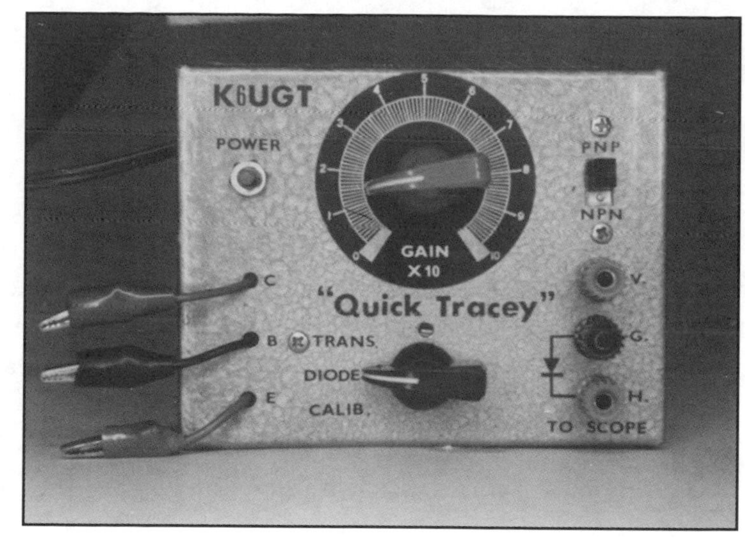

Figure 18-5: Exterior view of the completed unit.

Various transistor sockets could be wired in parallel instead of the three alligator clip leads. A screw-type terminal strip could replace the 5-way binding posts. Any 6.3 volt transformer will do; we used the least expensive one we could find. If you can't find a 6.3 volt transformer, use half of a 12 volt center-tapped transformer, such as a Radio Shack #273-1365. Function switch S2 is any 2-pole 3-position type, rotary or slide; you can use three positions of Radio Shack's #275-1386 2-pole 6-position rotary switch. The battery, a standard transistor radio 9 volt type, is held to the case with a home-made bracket made from a thin 1.5- by 2-inch piece of scrap aluminum.

The battery connector was salvaged from a dead battery of the same type. A terminal strip was used to anchor the alligator clip leads, and another terminal strip to anchor the power cord and transformer input leads. With the push-to-test power switch, a pilot light is not necessary. Figure 18-6 shows the interior of my unit; as you can see, there is actually plenty of room inside the cabinet for the parts and wiring.

Dry transfer labels or decals complete the job. You may have a problem finding the nice dial plate shown in the photo (30 years ago they were readily available), but you can instead easily mark the front panel with decal numbers or make a paper dial plate.

Figure 18-6: Interior of "Quick Tracey" showing parts placement and wiring.

Calibration

Regardless of whether you're "investigating" diodes, transistors, or whatever, you must first calibrate the unit. Plug the line cord into a 117 V 60 cycle source and connect the scope as shown in Figure 18-2. Be sure to set the scope sweep selector to the horizontal input. Put S2 in the "Calibrate" position and depress power switch S1. A slanted line will appear on the scope screen; adjust the vertical and horizontal gain and position controls until this line is in the center of the screen, at a 45 degree angle, and filling about two-thirds of the screen diameter. You are now calibrated for equal vertical and horizontal voltages at the Quick Tracey output terminals. As mentioned earlier, the line may slant in the opposite direction, in which case all other test traces will be flipped horizontally.

Sleuthing with Quick Tracey

DIODES: To test a diode or rectifier, connect it between terminals G (ground) and H (horizontal), with the cathode connected to H, as shown in Figure 18-3. Put S2 in the "Diode" position. The trace tells the story when Sl is depressed. Compare your pattern with the typical traces shown in Figure 18-3. Connecting the diode backwards will give you an inverted trace, which allows you to determine the cathode of unmarked diodes. Zener breakdown voltage can be estimated as a proportion of 10 volts by measuring the distance of the breakdown point from the junction, as compared to a regular diode (Figure 18-3 Zener pattern). Shorted or open diodes are instantly spotted by the straight vertical or horizontal line. Selenium rectifiers usually have a rounded junction and a slanted, shorter vertical leg. Tunnel diodes (where as the voltage increases the resistance increases, then decreases, then increases again) show a distinctive double-curve trace.

TRANSISTORS: Connect the collector, base and emitter transistor leads to the C, B, and E of Quick Tracey. Put the polarity switch S3 in the more common NPN position, unless you know for sure that you are testing an PNP transistor. Put function switch S2 in the "Transistor" position. When S1 is depressed you should get a sloping pattern with a flat bottom section. Adjusting Quick Tracey's gain control, R4, should change the slope. Don't touch the scope controls which you previously calibrated for equal gain! If you get no significant pattern, or the gain control has no effect, you may have a PNP transistor under test, so flick S3 to the PNP position. Still nothing? Make sure you have not misidentified or misconnected the leads.

On a good transistor, you will get a trace like the patterns shown in Figure 18-4. Using the gain control on Quick Tracey, set the slope to about 45 degrees, and read off the approximate gain on the R4 scale. Even with the gain control set to minimum, you still have a gain of 10, because of the series current-limiting resistor R3.

Of course, all this assumes that the transistor is properly connected. Transistor basing is pretty well standardized these days, and there are many sources of basing diagrams. If you're not sure, try various combinations; Quick Tracy is very forgiving of "goofs" and I have yet to hurt a transistor or diode by hooking it up wrong on Quick Tracey. You will get some mighty weird patterns with some misconnections, and that should tip you off.

Minor variations in the trace can be significant. For instance, if the sloped line is perfectly straight, the transistor has linear response (at least in the low current range). A curved sloping line is characteristic of RF transistors, which need not be linear in most applications. A short "tail" at the top of the NPN slope (or bottom of the PNP slope) is "leakage," which is probably not going to bother you unless you have a critical application. This tail is common, but no tail is preferable.

Incidentally, on all Quick Tracey testing, don't be upset if the traces show dual lines on some oscilloscopes. This is due to a non-linear condition in the scope common at the Quick Tracey working frequency of 60 cycles, and is not the fault of the part you are testing.

Defective transistors will either exhibit no trace at all, or one which is obviously not right. Finding defective transistors in "bargain" packs is a cinch with Quick Tracey. You can sort them out by approximate gain, linearity, and type (PNP or NPN) and use colored dots of paint for coding. Quick Tracey will uncover the high percentage of undesirables in many bargain packs.

Odd Jobs for Quick Tracey

Quick Tracey can be used for many other tasks, like these:

(1) *Relative Resistance Measurement.* Set for diode test. Connect unknown resistor as for diode, except there is no polarity consideration. A horizontal trace means a high resistance; as resistance decreases the trace slants more and more vertical. Vertical, as you recall, means a short circuit. A resistance of 100K shows an almost horizontal trace, tilting upward to about 20 degrees at 10 kilohms, 70 degrees at 1 kilohm, and essentially vertical at 100 ohms or less. This is a good way to test potentiometers for open spots or noise (noise will make the trace fuzzy as you rotate the shaft).

(2) *Capacitance Testing.* You can not only tell if the capacitor is good (at low voltage), but you can estimate the value of capacitance for all units from 0.05 microfarads to several hundred microfarads, including those difficult-to-test low voltage transistor electrolytics! Even more surprising, you don't have to worry about polarity when testing an electrolytic; just use Quick Tracey as for diodes. The pattern will be a horizontal long and thin ellipse for 0.05 mF, growing to a circle at about 1.0 mF, and becoming a vertical

ellipse beyond that value, slowly closing to a vertical line at several hundred microfarads (which is essentially a short circuit to 60 cycles). You can make a calibration chart from known values, plotting value against ellipse proportions.

(3) *Testing Photoconductors.* These devices have a very high "dark resistance," and a relatively low resistance when exposed to light. Connect the leads of a photoconductor as described for diode testing, except there is no polarity to worry about. Cover the face of the cell with your hand. When S1 is depressed you should get an almost horizontal line (depending on the normal dark resistance of the type of cell you are using). When you remove your hand and expose the cell to light, the line will tilt toward vertical if the cell is good. The more light, the more vertical. A graph of the trace angle plotted against light intensity could be used as a rough light measurement device. Some types of cells show relatively little change; others will go from straight horizontal to straight vertical!

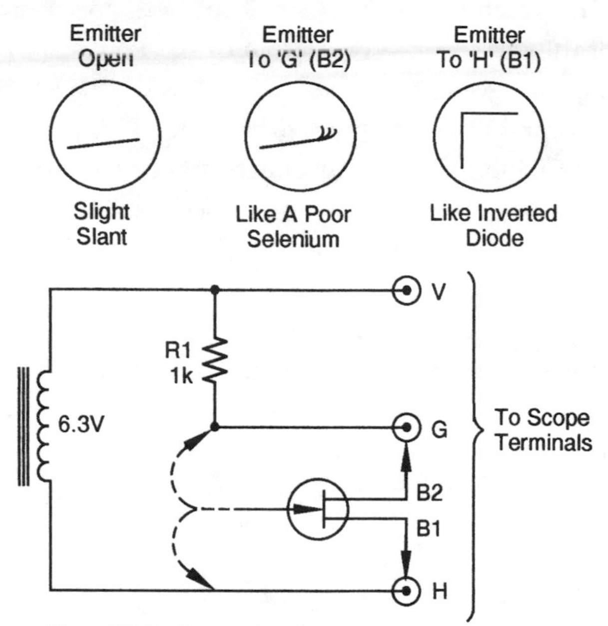

Figure 18-7: Connections for testing N-type unijunction transistors.

(4) *N-Type Unijunction Transistors.* As shown in Figure 18-7, connect Base 2 to the G terminal of Quick Tracey, Base 1 to the H terminal. Leave the emitter unconnected. Set S2 to "Diode" and depress S1. You should get a slightly slanted horizontal line, since the unijunction has a high resistance with an open emitter. Now touch the emitter to G and then to H and you should get the traces shown in Figure 18-7.

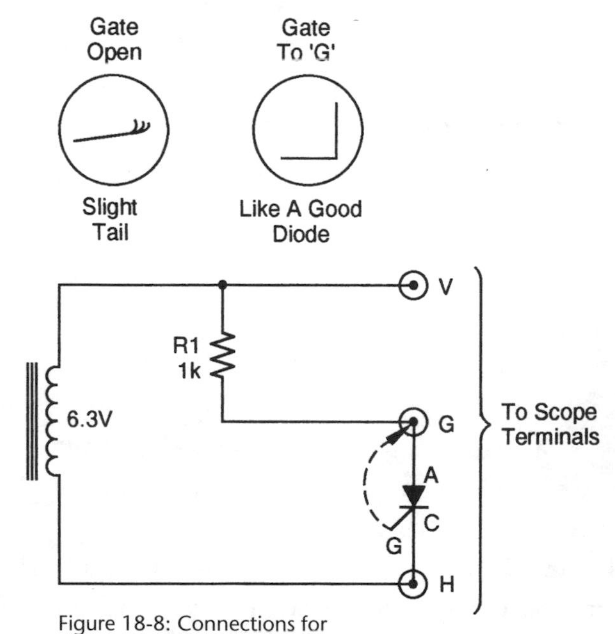

Figure 18-8: Connections for testing silicon-controlled rectifiers.

(5) *Silicon-Controlled Rectifiers (SCRs).* Connect the anode and cathode as shown in Figure 5B. Leave the gate unconnected. Set Quick Tracey for diode test. When S1 is depressed, you'll get a horizontal line, perhaps with a curved tail. When the gate is connected to G (use a clip lead) you'll get a trace that looks like a normal diode. The vertical leg shows that the SCR is properly conducting during the half-cycle when the anode and gate are positive with respect to the cathode.

My original Quick Tracey prototype has been in use for over 30 years, and has easily paid for itself in savings on bargain transistors, diodes and SCRs. Super-sleuth Quick Tracey will uncover defects that other testers miss, and show some things that only the most expensive testers check. Quick Tracey's fee is low for the service performed. Quick Tracey's motto: "Stamp out bad semiconductors!"

PARTS LIST

T1:	Small 6.3 volt filament transformer (see text)
S1:	Normally open SPST pushbutton switch (see text)
S2:	2-pole 3-position miniature rotary switch (see text)
S3:	DPDT slide switch
BP1, BP2, BP3:	5-way binding posts (Radio Shack 274-662)
B1:	9 volt battery, Burgess 2U6 or equivalent
R1, R2:	1 kilohm 1/4 watt carbon resistor
R3:	10 kilohm 1/4 watt carbon resistor
R4:	100 kilohm linear-taper potentiometer
Cabinet:	See text
Dial plate:	Mallory #380, Ohmite #5000 (see text)
Miscellaneous:	AC line cord, pointer knob, three alligator clips, dry transfer lettering, wire, solder, screws, nuts, aluminum scrap for battery mounting bracket, terminal strips.

CHAPTER 19

Digital Voltmeter Capacitance Adapter

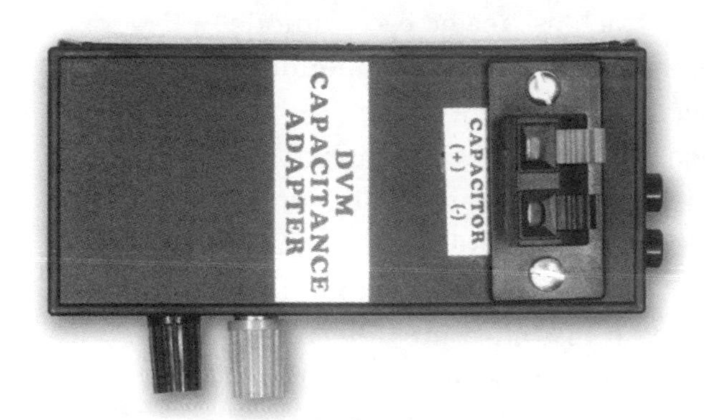

Capacitors come in many types, sizes, shapes, and an enormous range of typical values from several picofarads (pF) to thousands of microfarads (μF). This inexpensive capacitance adapter lets you use a common digital voltmeter (DVM) or digital multimeter (DMM) to measure capacitor values from 20 pF to 2 μF. It can be built from scratch or from a parts kit described at the end of this chapter.

Next to resistors, capacitors are the most common components in typical circuits. Capacitors come in an amazing number of sizes and shapes, and with an enormous range of values.

The many types of capacitors are generally based on the material used for the dielectric. There are ceramic capacitors, mica capacitors, paper capacitors, synthetic film capacitors, electrolytic capacitors (polarized and non-polarized), tantalum capacitors, and variable capacitors.

Typical capacitance can range from a few picofarads (one millionth of a microfarad) to thousands of microfarads (one millionth of a farad.) To add to the confusion, capacitor values are identified in different ways, and their actual values are frequently far different from their marked values.

Value markings on capacitors are not nearly as standardized as on resistors. Many, especially those used in some commercial products, have no markings at all! Those that are marked may use simple numbers, coded numbers, color coded dots, or colored stripes. Some have tolerance codes and temperature limit codes. However, unless a capacitor is clearly marked with its nominal value, or uses a three-digit numerical code that is somewhat similar to the resistor multiplier code, you could have no idea of its capacitance value or its voltage limitation. If you really need to know the actual capacitance, a proper measuring instrument is needed.

However, measuring the actual value of a capacitor is not anywhere near as simple as measuring a resistor value. Digital multimeters apply a constant current through a resistor and measure the voltage across it. (Ohm's law: Resistance in ohms is equal to the voltage in volts divided by the current in amperes.) The greater the voltage at a set current, the higher the meter reading. Appropriate meter markings or calibration make measuring resistance a relatively straightforward task.

But measuring the actual capacitance value of an unknown capacitor takes some form of dynamic testing, since capacitors are initially not charged and their "capacity" is what must be measured. Capacitors are inherently AC devices, since they block DC current flow once they are charged, and the amount of AC current they pass is proportional to the AC frequency.

Many methods have been devised to measure the value of capacitors, from various AC bridges with null indicators (such as "magic eye" cathode ray tubes or zero-center meters) to oscillators that measure the amount of current flowing through a capacitor at a specified frequency.

I've owned a "Lafayette Condenser Checker" (Model LC-15) since 1959 (we used to call capacitors "condensers" in those days) and it still works fine. It uses an AC bridge circuit with a 6E5 magic eye null indicator, and covers a test range from 10 picofarads (pF) to 1000 microfarads (µF) in four overlapping ranges with about 10% accuracy. It also measures the power factor of electrolytic capacitors, and provides five voltage test ranges up to 450 volts. I believe it cost me about $25 back in 1959.

But today you can't find simple instruments like these. Lafayette, Eico, Allied, Heath and others are either out of business or no longer make simple equipment of this sort even in kit form. (Ah, for the "good old days"!)

However, you can measure from 20 pF to 2 µF with a digital voltmeter or digital multimeter you probably already have, using an adapter circuit. This chapter will describe the construction of such an adapter from scratch or using a parts kit. Incidentally, the adapter described in this chapter will not give proper readings with an analog meter; you need a digital meter with its typical 10 megohm input impedance. And this adapter does not reliably measure capacitors higher than 2 µF, so it does not test most electrolytics. It does not perform any voltage tests, nor measure the power factor of electrolytics.

Circuit Description

Figure 19-1 shows the schematic of the CA-1 Capacitance Adapter. The LM7805 voltage regulator (U2), together with capacitors C3 and C4, provides a stable 5 volts DC even as the 9 volt battery drops in voltage over time. As long as the battery supplies about 7 volts, the regulator will provide 5 volts to the rest of the circuitry. Since the CA-1 current drain is only about 5 milliamperes, and there is an on/off switch, the battery can last a very long time with its typical occasional use.

Integrated circuit U1 is a 74HC132 quad NAND Schmitt trigger, which means it has four fast switching NAND gates. Each gate is independent of the others. J1 and J2 are the input points for the unknown capacitor.

Analyzing the operation of this circuit is very tricky. Begin by assuming no capacitance at J1-J2, and the double-pole double-throw range switch,

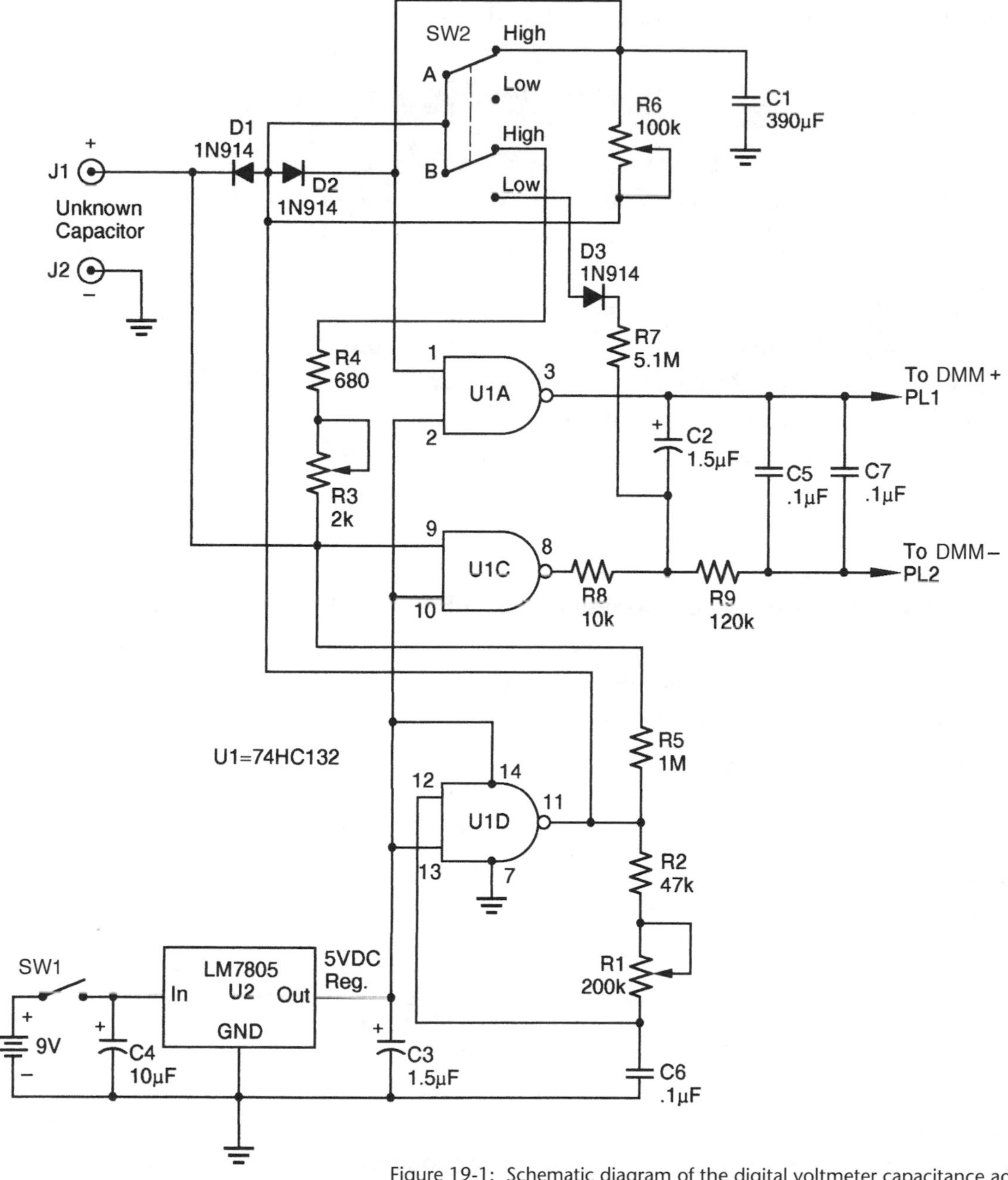

Figure 19-1: Schematic diagram of the digital voltmeter capacitance adapter.

SW2, is in the LOW position. When power switch SW1 is closed, battery voltage is regulated to 5 volts by U2 and this 5 volts (which we'll call "high") is applied to one input of each of the three NAND gates used—pin 2 of U1A, pin 10 of U1C, and pin 13 of U1D. (U1B is not used.)

When a NAND gate has any combination but high on both inputs, the output is high, so the outputs of all three NAND gates are initially high. However, look at pin 11 of U1D. The high here is applied through resistor R2 and potentiometer R1 to capacitor C6, allowing C6 to charge at a rate based on the setting of R1. This is called an RC (resistance-capacitance) circuit, where the capacitor charging time is a function of the resistance and capacitance values; the higher either value, the longer the charging time.

C6 is directly connected to the other input, pin 12, of U1D. When C6 reaches the "trigger" voltage of U1D (about 2.5 volts), U1D sees both inputs as high, and the output, pin 11, snaps to a low (ground) voltage. But now C6 can discharge through R1 and R2 to pin 11, so C6's voltage drops. When it gets down to about 1.5 volts, pin 12 drops low so pin 11 jumps back to high, and the cycle repeats. This is an oscillator, and with the values used it switches about 160 times a second, producing a non-symmetrical 5-volt square wave at pin 11.

Meanwhile, input pin 9 of U1C (through resistor R5) and pin 1 of U1A (through diode D2) follow pin 11, producing square wave outputs at pins 3 of U1A and 8 of U1C. Potentiometer R6, together with capacitor C1, steering diode D3, and resistor R7, allow adjustment of the square wave at pin 3 to be identical to the square wave at pin 8. This is the "zero-adjust," so there is no voltage between output terminals PL1 and PL2 going to the external digital voltmeter.

So with no capacitor at J1-J2, the digital voltmeter reads "000" when potentiometer R6 is properly adjusted. Now assume a 0.001 μF (1000 pF) capacitor is placed between J1 and J2. Pin 11 oscillates between high and low at a rate adjusted by the setting of R1. But now when pin 11 is high it is charging the 0.001 μF test capacitor, with some delay in charging and discharging due to resistor R5. This means the square wave output at pin 8 will be low longer than it is high, while U1A pin 3 continues high longer than it is low. It is this difference in the duration of the high at pin 3 and the shorter high at pin 8 that creates a positive voltage which is filtered by C2, C5, C7, R8, and R9 so that a steady DC voltage appears at PL1 and PL2. In other words, the circuit is now measuring "pulse width difference."

LOW switch calibration is obtained by setting potentiometer R1 so that the digital voltmeter reads 1.0 volts when a .001 μF capacitor is used at J1-J2. With a smaller value of capacitor under test, the square wave at pin 8 is low for less time, so the pulse width difference between pins 8 and 3 is less, and the meter reads less.

If switch SW2 is put in the HIGH position, resistor R4 and potentiometer R3 affect the timing of the high at pin 9 of U1C, altering the square wave output at pin 8. This compensates for higher unknown capacitor values. However, the same pulse width comparison takes place between pins 3 and 8, yielding a positive voltage value between PL1 and PL2, adjusted by R3 for 1.0 volt DC at PL1-PL2 with a 1.0 μF capacitor at the JI-J2 input.

Assembly

Figure 19-2 shows a printed circuit board layout for the adapter, and Figure 19-3 shows the parts layout using this PC board. You can build the adapter from scratch with parts you already have or you can buy from various vendors. However, if you do this, you may have difficulty fitting the parts (especially the switches and potentiometers) to the printed circuit layout shown. If you buy the parts kit described later in this chapter, all parts, including the printed circuit board and a terminal strip for the unknown capacitor, are included.

If you build the adapter from the kit, assembly is relatively fast and easy. Mount all the close-to-the-board parts first, then the switches and potentiometers. Be careful that the three diodes, capacitors C2, C3, and C4, and U2 are properly oriented as shown in Figure 19-3. Don't forget to add the jumper shown.

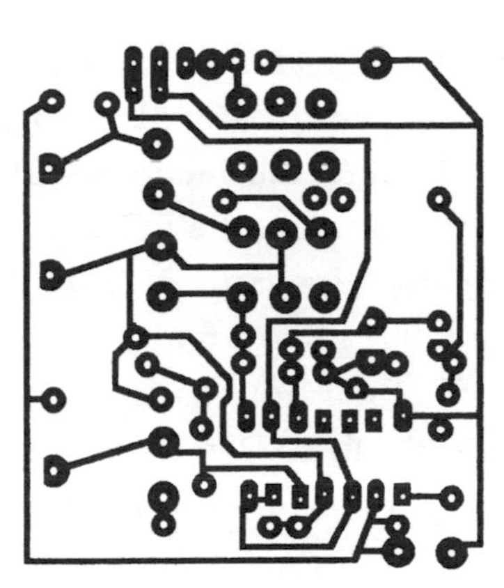

Figure 19-2: Printed circuit board pattern (actual size).

Figure 19-3: Parts layout and placement for the capacitance adapter.

Use a socket for U1 (supplied in the parts kit) and be sure to install pin 1 on the right as shown. Connect the battery snap (supplied), making sure the red wire goes to the (+) shown in Figure 19-3, and the black to (-). Use some hook-up wire to connect the unknown (+) capacitor input to the red terminal of the supplied input jack, and (-) to the black jack terminal. Similarly, connect two wires to the (+) and (-) DVM outputs to connect to your digital voltmeter, marking the positive wire.

Figure 19-4 shows the circuit board with all parts mounted and installed.

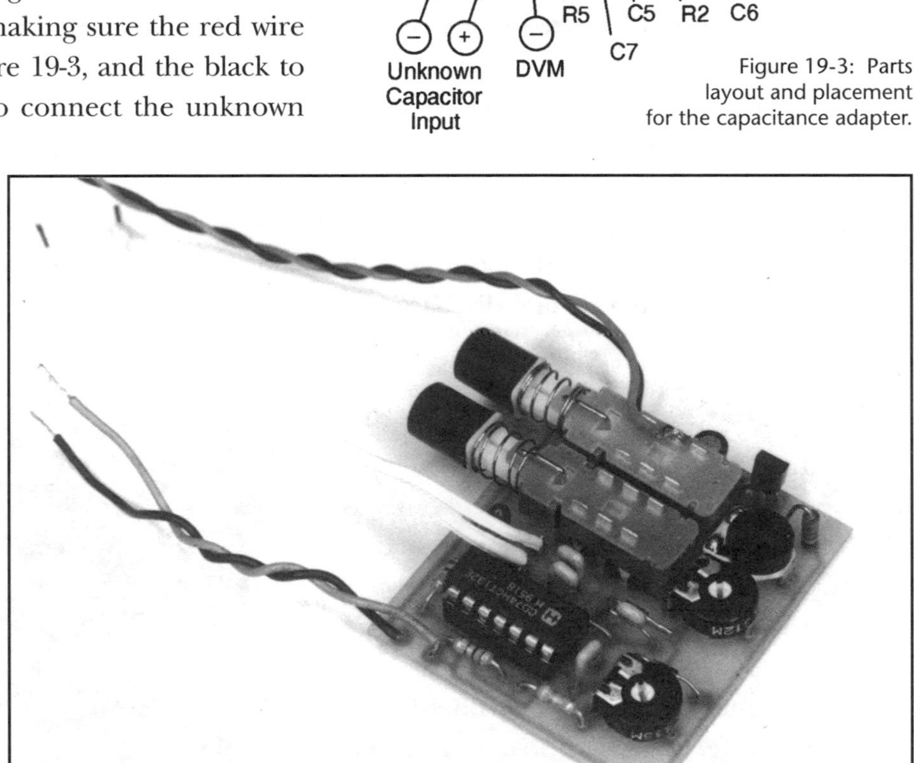

Figure 19-4: Circuit board with all components installed.

Testing

The toughest part of the testing is finding the 0.001 μF and 1.0 μF "calibration" capacitors. (These should really be provided with the parts kit.) Since capacitors vary notoriously from their markings, try to use precision capacitors, or find the actual values using another capacitance tester and adjust your calibration accordingly.

With no capacitor connected to the input terminal J1 and J2, and range switch SW2 in LOW, connect the battery, and press SW1 to power the unit. With the digital voltmeter attached to PL1 and PL2, and set to the 2 volt range, you should see a reading of close to zero. Adjust R6 until the reading is zero.

Now place the "known" 0.001 µF capacitor at J1-J2. The meter should now read about 1.0 volts. Adjust R1 until it does. Next use the 1.0 µF calibration capacitor at J1-J2, switch the range to HIGH, and the meter should read around 1.0 volts. Adjust R3 until it does. I've found that the LOW position of the range switch should be used for capacitor measurement from 20 pF to 0.001 µF, and the HIGH switch position for 0.001 µF to 2.0 µF.

Don't forget to turn the power switch to "off" when not in use, or the battery will drain. You might want to consider adding an LED and 1 kilohm resistor as an "on" indicator. Just connect them in series between switched voltage and ground, observing LED polarity.

Troubleshooting

If your adapter doesn't work, first verify that all parts are in the right place, are the proper values, and are oriented as shown. Make sure the solder joints look good: not gray or grainy.

Next check the voltages at U1, looking for a steady 5 volts on pins 2, 10, 13, and 14, and 0 volts at pin 7.

Beyond that, you'll need an oscilloscope. Look for square waves on pins 3, 8, and 11 referred to ground, and then check for a square wave between pins 3 and 8. Also, pin 12 referred to ground should show a 2.5 volt peak-to-peak triangle wave as C6 charges and discharges.

Packaging

I've used a black "Fuji Slide Box" for many of the small projects in this book. This is normally used by film finishers to hold 36 color slides. It is made of soft plastic, easily drilled, and cut. See if you can get some from your photo finisher.

The adapter printed circuit board and battery fit perfectly in this box, as you can see in Figure 19-5. I added two binding posts to allow easy connection to my digital voltmeter leads. The two switch knobs fit through round holes in one end of the box, and the unknown capacitor terminal mounts on the top. Figure 19-6 shows an external view of the completed unit.

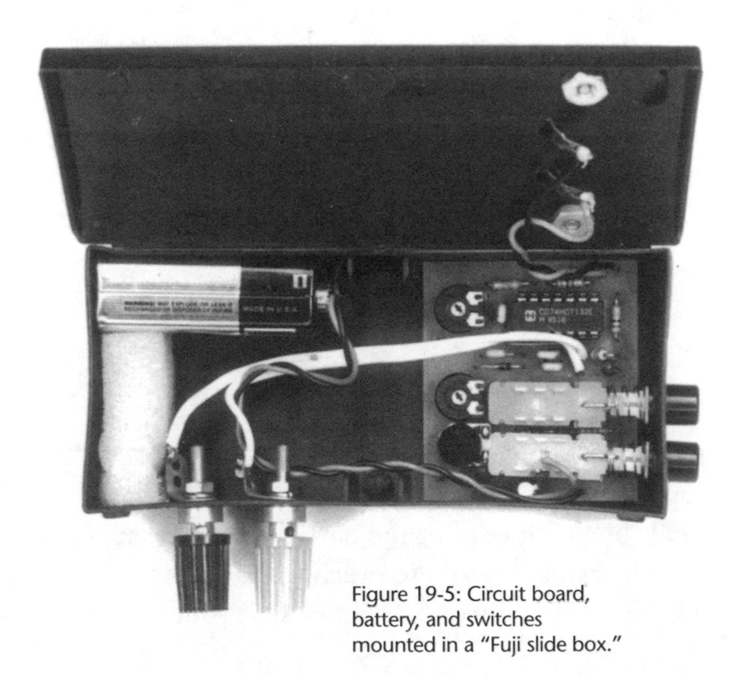

Figure 19-5: Circuit board, battery, and switches mounted in a "Fuji slide box."

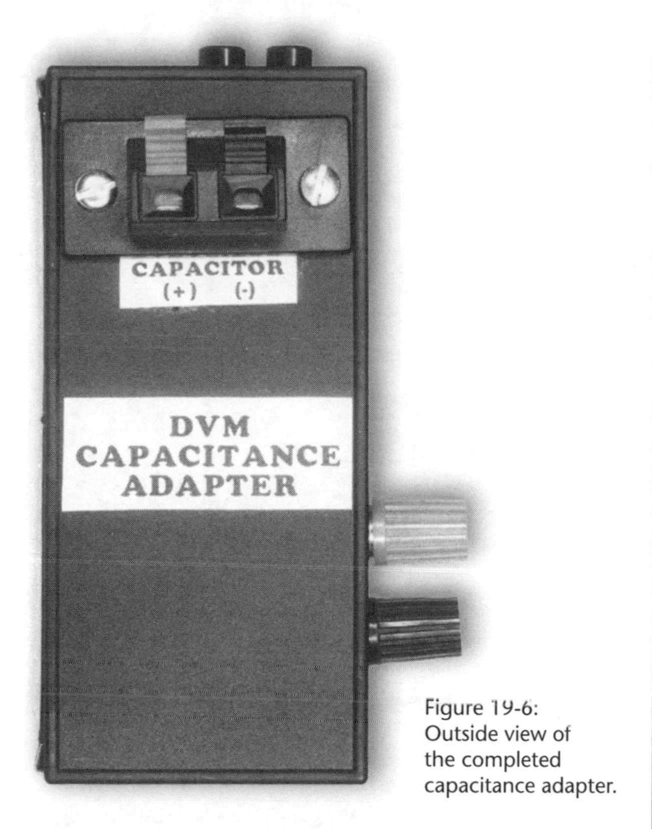

Figure 19-6: Outside view of the completed capacitance adapter.

A complete kit of all parts listed above, except for the wire, solder, cabinet, and battery, is available from Electronic Rainbow, Inc., 6227 Coffman Road, Indianapolis, IN 46268; telephone (317) 291-7262, fax (317) 291-7269. Order the "CA-1 Capacitance Adapter." Price at the time this was written was $12.95 each plus $5 shipping and handling (Canada $7 shipping and handling). Pricing and availability may have changed since this book was published.

PARTS LIST

C1:	390 pF monolithic ceramic or mica capacitor
C2, C3:	1.5 µF tantalum capacitor
C4:	10 µF electrolytic capacitor
C5, C6, C7:	0.1 µF monolithic ceramic capacitor
D1, D2, D3:	1N914 or 1N4148 signal diode
J1, J2:	Dual input jack
J3:	9 volt battery snap connector
R1:	200 kilohm trim potentiometer
R2:	47 kilohm 1/8 watt 10% carbon resistor
R3:	2 kilohm trim potentiometer
R4:	680 Ω 1/8 watt 10% carbon resistor
R5:	1 megohm 1/8 watt 10% carbon resistor
R6:	100 kilohm trim potentiometer
R7:	5.1 megohm 1/8 watt 10% carbon resistor
R8:	10 kilohm 1/8 watt 10% carbon resistor
R9:	120 kilohm 1/8 watt 10% carbon resistor
SW1, SW2:	DPDT printed circuit mount push switch
U1:	74HC132 quad Schmitt trigger integrated circuit
U2:	LM7805 or LM78L05 5 volt DC regulator
Miscellaneous:	14-pin integrated circuit socket, wire, solder, cabinet, 9 volt battery.

CHAPTER 20

A Function Generator

There are many specialized signal generators available, some costing hundreds of dollars, to produce input signals of different frequencies and waveforms. The circuit described in this chapter will produce three different types of waveforms—sine, square, or triangle—from 1 Hz to above 250 kHz with adjustable voltage output for the sine and triangle waveforms. You can build this function generator from parts you might already have or from a parts kit.

For simple "static" (nothing changing rapidly) testing, probably the most common test instrument is a multimeter. This will measure current, voltage, resistance, and sometimes capacitance.

If the circuit is operating under "dynamic" (fast changing) conditions, you really need a "signal tracer" to see where the desired signal is getting blocked. A typical signal tracer would be an oscilloscope, amplifier, speaker, or specialized detector. The more specialized the circuit, the more sophisticated a device might be required as a tracer. For example, television or microwave troubleshooting requires more specialized equipment than an audio amplifier.

With circuits that create their own dynamic action, such as oscillators, no other input is needed. But for amplifiers and many other electronic devices, you need a device to produce an appropriate input signal for tracing. This is generally known as a *signal injector, signal generator,* or *function generator.*

This chapter describes how to build a function generator. This circuit will produce sine waves, square waves, or triangle waves, as shown in Figure 20-1. A selector switch chooses three different frequency ranges from 1 Hz to above 250 kHz. Although this primary frequency limit is below broadcast frequencies, square waves can be useful for testing AM and even FM radio circuits since they consist of many harmonics.

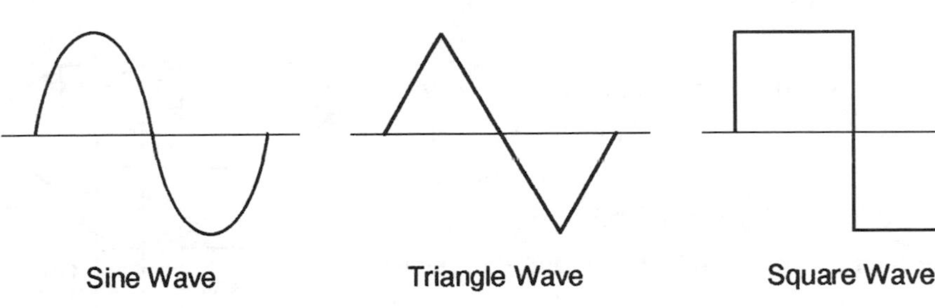

Figure 20-1: Sine, triangle, and square wave forms.

Circuit Description

Figure 20-2 shows the schematic of the function generator. The heart of the circuit is an XR-2206 monolithic function generator integrated circuit, which can generate sine, square, triangle, ramp and pulse waveforms. A pin-out and functional block diagram of the XR-2206 is shown in Figure 20-3. In this project, only the sine, triangle, and square waves are used.

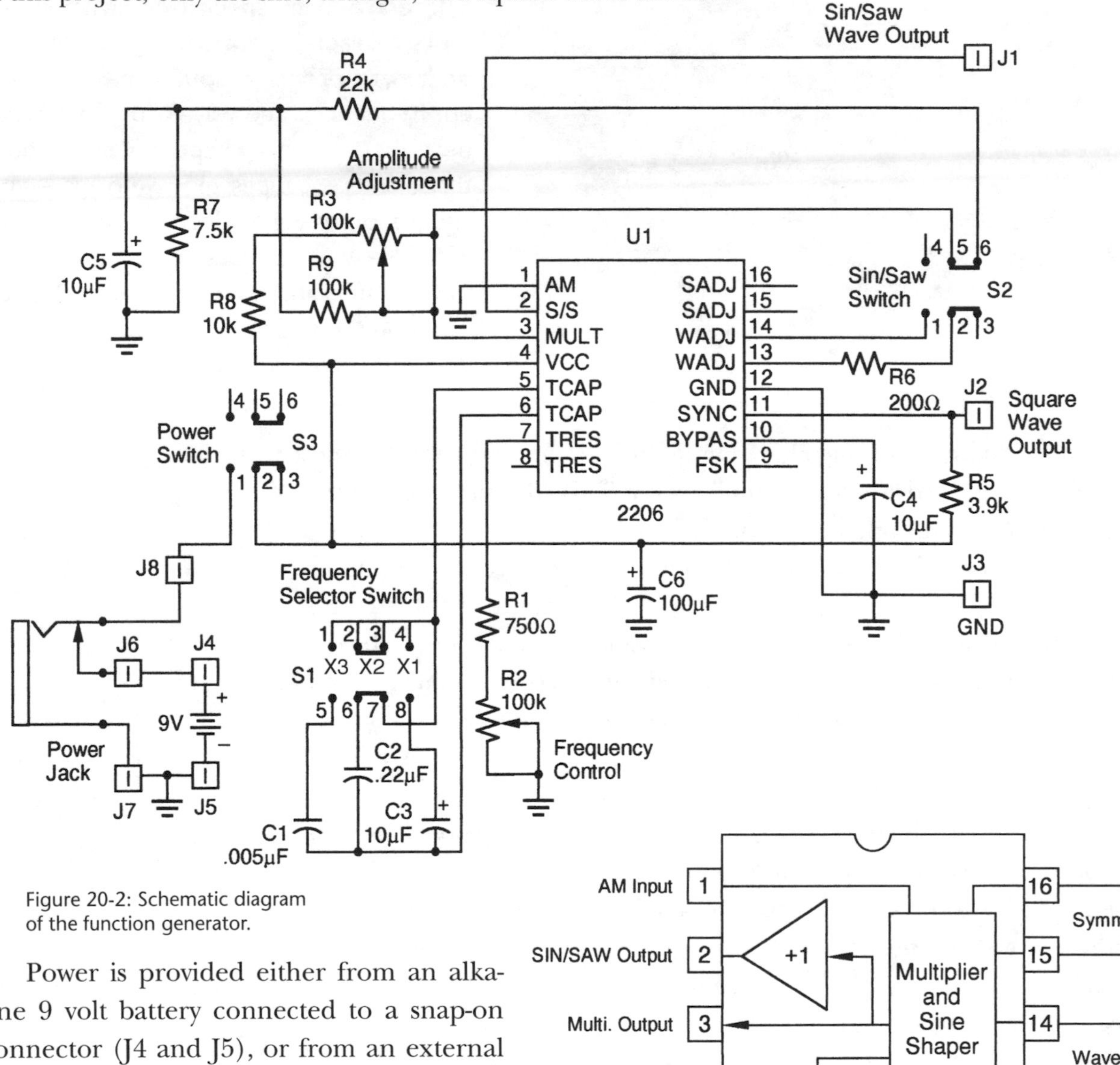

Figure 20-2: Schematic diagram of the function generator.

Power is provided either from an alkaline 9 volt battery connected to a snap-on connector (J4 and J5), or from an external DC power source of 9 to 18 volts DC into a power jack. When the external power plug is inserted into the power jack, the battery is disconnected. To assure that a typical DC wall transformer (which might be unregulated and have significant "ripple voltage") can be used, a high-value electrolytic capacitor (C6) is used to filter the ripple.

The frequency of operation is determined by the value of the capacitor selected

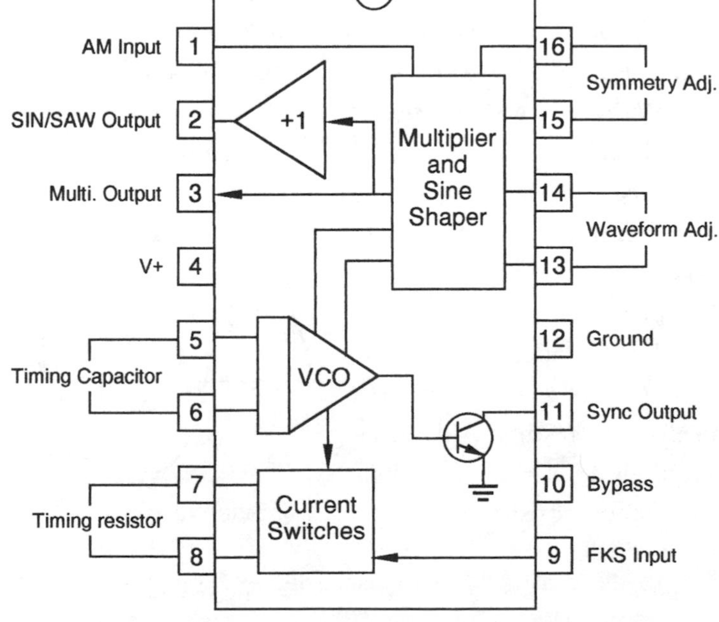

Figure 20-3: Internal diagram of the XR-2206 IC.

by switch S1, and the setting of potentiometer R2. Should you wish to set your own frequency ranges, the formula for determining the frequency is F = (1/RC), where F is the output frequency in Hertz, R is the total resistance of resistor R1 and the setting of potentiometer R2 in megohms, and C is the actual value (not necessarily the marked value) of the chosen capacitor, C1 or C2 or C3, in microfarads. Note that the frequency is inversely proportional to the value of R times C; that is, the higher the value of either R or C, the lower the frequency.

Here are the calculated maximum and minimum frequencies for each of the three settings of switch S1:

Switch Setting	*Minimum Frequency (Hz)*	*Maximum Frequency (Hz)*
X1	1	133
X100	45	6060
X10K	2000	270,000

Bear in mind that typical component values are usually considerably higher or lower than the marked value, especially for capacitors, and even more especially for electrolytic capacitors!

Switch S2 selects either the sine or triangle wave shape that appears at output terminal J1. When resistor R6 is in the circuit, a sine wave is produced; when no resistor is in the circuit, the output is a triangle wave. In both cases, potentiometer R3 adjusts the amplitude of the output. The square wave has no amplitude adjustment, and appears at terminal J2.

Construction

You have several choices in deciding how to build this kit. You could simply buy the complete parts kit described at the end of this chapter. This includes all electronic parts and hardware, a drilled and silk-screened printed circuit board, a cabinet with a punched and silk-screened panel, hookup wire, solder, and a liberally illustrated assembly and instruction manual. The result is a professional looking instrument, with no hunting for odd parts.

You could also use parts you already have, or can get locally, and just purchase the special parts from the source given in the parts list. However, by the time you get the individual parts, (especially if you order the PC board, case, top panel, and integrated circuit chip) it will probably cost you more than the parts kit.

(NOTE: If you build this function generator from the parts kit, the front panel and instruction manual use the term "saw wave" in several places as a contraction of the proper term, "sawtooth wave." In any case, this is incorrect, since the signal produced is a "triangle wave" with equal rise and fall times. A true sawtooth waveform has a relatively slow rise time and a fast fall time. I will refer to this signal as a triangle wave for the remainder of this chapter.)

The printed circuit board traces for this project are shown in Figure 20-4. Notice that no holes are shown, since the parts you use may require different sized holes than those used with the kit parts. Parts locations, using the kit parts, are as shown in Figure 20-5. The kit PC board, available separately (see parts list), comes with all holes drilled, and the part locations silk-screened on the component side.

If you provide your own parts, and make your own PC board, you may have to adjust some part locations, depending on the size and shape of the parts you use. For example, you may use different sized potentiometers or switches than those provided with the kit, and they may be mounted differently. Here, again, I wish to emphasize that it is far easier, unless you're an experienced electronic project builder, to buy the parts kit.

In constructing this project, only a few of the parts have a specific polarity or orientation. When installing the electrolytic capacitors C3, C4, and C5 make sure the positive lead is installed with the proper polarity. Also be sure the integrated circuit notch, the Pin 1 end, is positioned as shown in the parts layout, Figure 20-5.

The subminiature power jack provided in the kit is apparently internally wired differently than those furnished previously. The red and green wires going between the PC board and the power jack upper terminals are shown incorrectly in the instruction manual. They should be reversed, as shown in a bulletin that accompanies recent kits.

The cabinet that comes with the kit has an internal wall at the bottom, intended to provide a compartment for the battery. If your battery won't fit (mine didn't), simply use a pair of pliers to break out the wall; it snaps off easily.

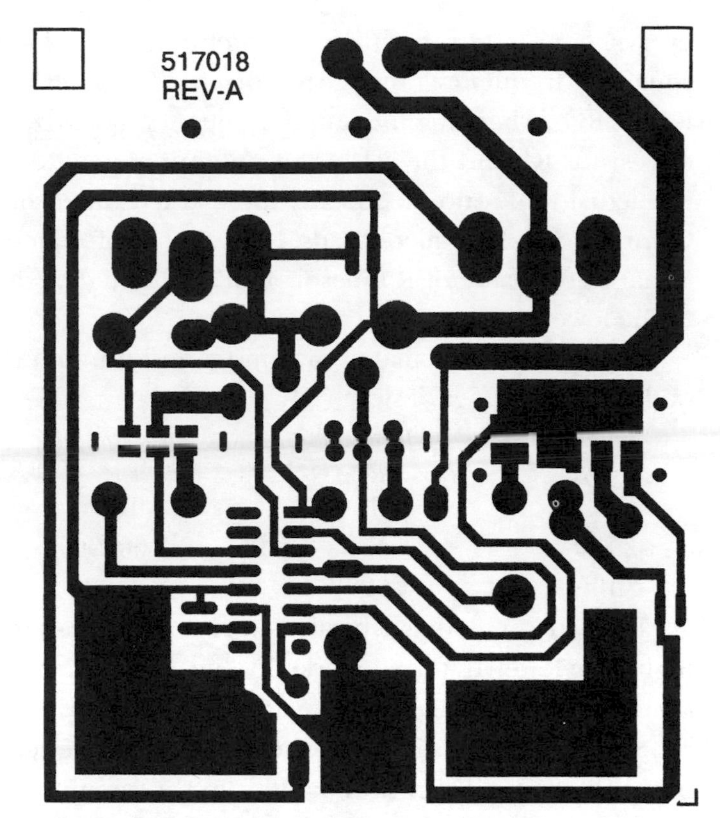

Figure 20-4: Pattern for the printed circuit board (actual size).

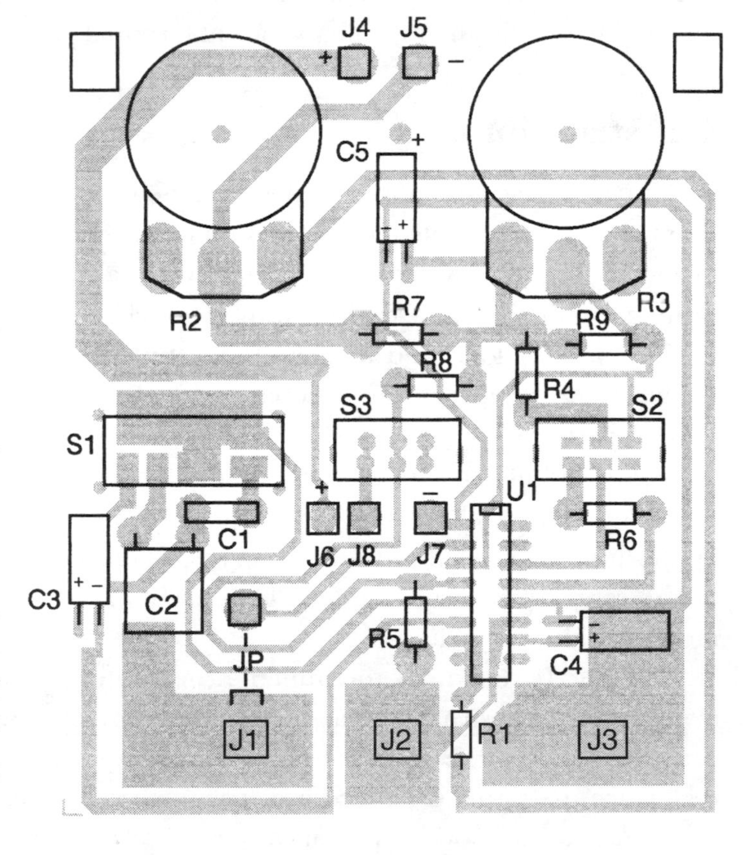

Figure 20-5: Parts location and placement on the PC board.

Figure 20-6 shows the assembled PC board that comes with the parts kit; note the generous amount of space that makes for easy construction. The pencil points to the XR-2206 integrated circuit.

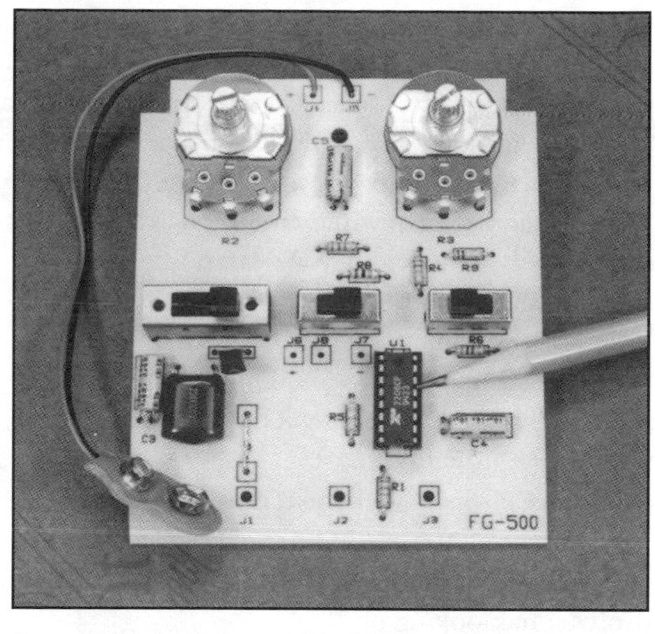

Figure 20-6: Assembled circuit board for the function generator; pencil points to the XR-2206 IC.

Modifications

If you always use a battery to power the circuit, then electrolytic capacitor C6 shown on the schematic is not required. It is not included in the kit. If you use a common wall-plug unregulated external power supply, it probably has a lot of "ripple" (AC voltage riding on the DC voltage), which will drastically impair performance, and C6 WILL be required. It can be from 100 to 1000 microfarads at 25 working volts and should be mounted on the foil (underside) of the PC board. The positive lead connects to a center terminal of the S3 power switch, and the negative lead is soldered to any nearby convenient ground trace. Figure 20-7 shows the location of this capacitor.

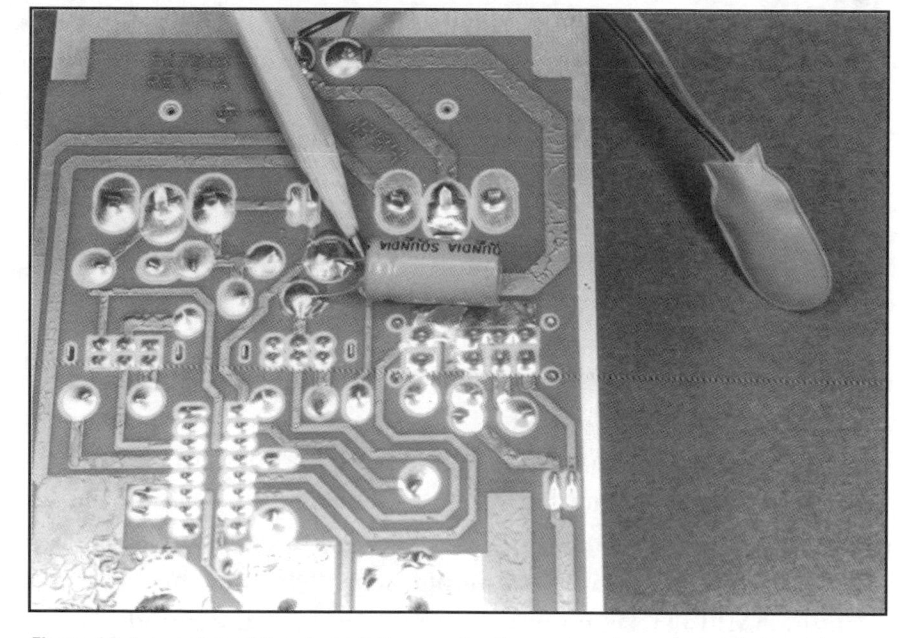

Figure 20-7: Location of the electrolytic capacitor if an external power supply is used.

While constructing the unit, either from scratch or from a kit, you might give some consideration to adding a small LED (light-emitting diode) in series with a 1000 ohm resistor to act as an ON light. Wire them between the ON terminal of the power switch S3 and ground, observing LED polarity, and placing the LED so it shows on the front panel. This is especially important when operating on battery, since there is no way to tell when the unit is on, and the battery draws considerable current when switch S3 is ON. Sure, the LED draws even MORE current, but it could save you from draining the battery when the Function Generator is not in use. When using external power the extra LED drain is not important, and you'll know when the unit is on.

You might also want to consider changing the subminiature power jack to a more rugged and common closed-circuit miniature jack, or a coaxial type of power jack.

Although some may consider this a small point, my opinion is that the power switch, S3, is "backwards." As provided on the printed circuit board (and shown on the schematic), you must push the switch to the LEFT to turn the unit on. It is far more conventional to move UP or RIGHT to turn on a switch. If you're building from scratch, this is easy to change, but it's not so easy using the PC board.

Figure 20-8 shows the exterior of a unit built using the parts kit. The screened faceplate of the cabinet makes for a very professional looking unit!

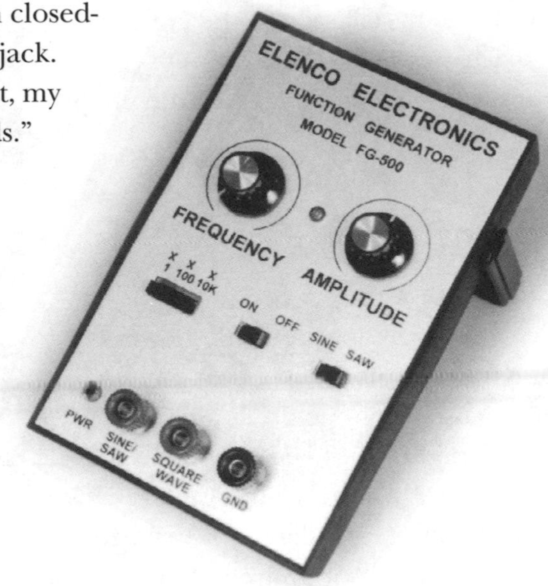

Figure 20-8: External view of the function generator built using the parts kit.

Testing

Although battery power is convenient, this unit uses a significant current (about 9 milliamperes without the optional LED, about 15 milliamperes with the LED.) If the voltage drops below about 7.5 volts, the sine and triangle waves start distorting. Battery life is relatively short, so if you intend to use this unit very much, plan on external power. And you might as well use external power when testing to assure that a low battery voltage does not give you false results.

Probably the easiest way to initially test the unit is with any small speaker. These usually have 4 to 10 ohm voice coils, but it won't make much difference if the test speaker has a higher value. Simply connect the sine/triangle output of the circuit to one voice coil lead, and the circuit ground post to the other voice coil lead. Set the ON-OFF switch to ON, set the frequency switch to X100, set the AMPLITUDE control to maximum, and as you turn the FREQUENCY control you should hear the sound through the speaker. Turn the AMPLITUDE control to see that it varies the sound intensity.

Now switch the frequency switch to X10K, and the sound should go up in pitch and vary as you turn the FREQUENCY control. Again check the AMPLITUDE control.

Next switch the "hot" (ungrounded) voice coil lead to the square wave output post of the FG-500. While the FREQUENCY control and switch still work, the AMPLITUDE control does not affect the square wave output.

If you have an oscilloscope handy, power the function generator at about 9 volts and you should get a sine wave or triangle wave of about 3 volts or greater peak-to-peak at maximum amplitude with the FREQUENCY control at maximum and the selector switch set at X1. The square wave output should be 5 volts or greater peak-to-peak.

Troubleshooting

Most problems encountered when troubleshooting items made from kits can be traced to poor soldering, wrong parts placement, or reversed installation of polarized parts.

After checking the soldering, if there is no sine/triangle or square wave output, first check that voltage is getting to and past the ON/OFF switch. Be sure the integrated circuit notch-end is properly placed, and that the electrolytic capacitors are not oriented backwards. The DC supply voltage should appear on pin 4 of IC1.

If you're getting the wrong frequency range settings, see that the capacitors are connected to the proper switch positions on S1. For any other malfunction, check the associated components for soldering and misplaced resistor values.

Using the Function Generator

Your use of this function generator will largely depend on the type of equipment you'll be testing. For analog equipment, such as audio amplifiers, this circuit provides a sine wave input signal in the audio range, with a speaker as your output "signal tracer." For digital circuits, use the square wave output, and the lowest frequency setting so you can see the effect of each signal transition.

For radio frequencies, this circuit offers enough strength and harmonics to allow you to clip the output lead from the square wave post directly to a receiver antenna to get a solid signal. Use the X100 or X10K positions of the frequency selector switch. You can move through a circuit stage by stage to determine proper operation.

You can check for distortion by using the function generator for sine or triangle wave input and monitoring the output of a circuit with an oscilloscope to see if it changes the wave shape.

A complete kit including all the above parts (except C6) (kit FG-500K) is available from C & S Sales, Inc., 150 W. Carpenter Ave., Wheeling, IL 60090; telephone (800) 292-7711. The cost for the parts kit at the time of this writing was $34.95 plus $5 shipping ($8 to Canada). Individual parts were also available; their numbers are included in the list above. Call for latest prices and availability of all components above, as changes may have been made since publication of this book.

PARTS LIST

(Note: all resistors are carbon, 5%, 1/4 watt)

R1:	750 Ω
R2, R3:	100 kilohm potentiometer (C & S #192612)
R4:	22 kilohm
R5:	3.9 kilohm
R6:	200 Ω
R7:	7.5 kilohm
R8:	10 kilohm
R9:	100 kilohm
C1:	0.005 µ F 100V
C2:	0.22 µ F 100V
C3, C4, C5:	10 µF electrolytic 16V
C6:	100 µF electrolytic 25V (see text)
U1:	XR-2206 integrated circuit (C & S #332206)
J1, J2:	Red binding post (C & S #625011)
J3:	Black binding post (C & S #625010)
S1 -	Double-pole 3-position slide switch (C & S 541203)
S2, S3:	DPDT slide switch (C & S #541009)
Socket:	16-pin integrated circuit (C & S #664016)
Power Jack:	Subminiature closed-circuit (C & S #622130)
PC board:	Drilled, silk-screened (C & S #517018)
Case:	Black plastic with handle (C & S 623015)
Case Top Panel:	Punched and silk-screened (C & S #614101)
Knobs:	2 (C & S #622009)
Battery Snap:	For 9 volt battery (C & S #590098)

Hardware, hookup wire, solder, etc.

CHAPTER 21

Measuring Inductance with "Quick Henry"

"Quick Henry" is a simple, inexpensive, easy-to-build adapter that is used with an audio frequency (AF) or radio frequency (RF) generator to determine the inductance value of many types of coils from values as little as one microhenry to beyond one henry! No printed circuit board is required to build this circuit.

Measuring the inductance of a choke, transformer, intermediate frequency (IF) can, toroid, open air, or any other type of coil is a relatively clumsy process for the home experimenter or small repair shop. The measurement of inductance is usually avoided because of the difficulty and expenses in making the measurement.

In laboratories, where precision is important, relatively expensive Maxwell or Hay bridge circuits are used for inductance measurements. Some of the newer digital multimeters measure inductance as an additional feature, although their inductance is less accurate.

However using "Quick Henry" and two simple charts, anyone with an audio frequency (AF) generator can measure inductance quickly and easily from 1 millihenry (mH) to over 1 henry (H). With the use of a radio frequency (RF) generator, measurements can be made down to one microhenry (μH). Furthermore, using a calculator or simple BASIC computer program, measurements can be calculated beyond these ranges.

If you have the use of an AF or RF generator, Quick Henry will also allow you to determine the resonant frequency and sharpness of response ("Q") of audio and RF circuits from below 100 cycles per second (hertz, or Hz) to over 15 megahertz (MHz) with reasonable accuracy. The resonant frequency of unmarked IF transformers can be easily identified, and you can determine the inductance of unmarked filter chokes, slug-tuned coils, RF chokes, etc. You can design, test and trim air-wound coils and audio band pass circuits to your requirements.

Quick Henry's built-in meter indicator is optional, and may be omitted by the builder if a 10,000 ohms/volt or better DC multimeter, 100 microamp DC meter, or oscilloscope is available as an external indicator.

Circuit Description

Figure 21-1 shows the schematic of Quick Henry. Assume an audio generator sine wave signal is connected to binding posts BP2 and BP3. The signal passes through isolation resistor R1 and then through the unknown inductance connected across BP4 and BP5. Switch Sl is used to select an appropriate value of capacitance (C2, C3, C4, or C5), which is placed in parallel with the unknown inductor.

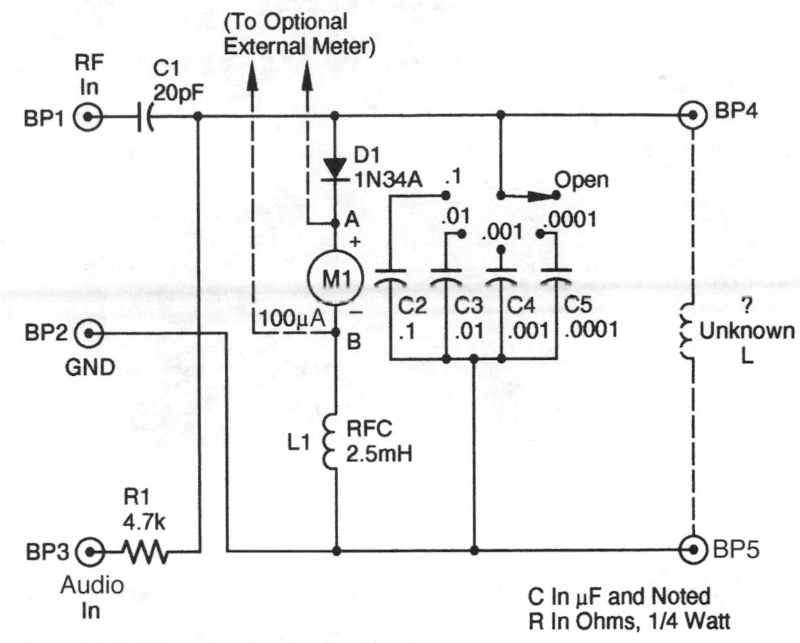

Figure 21-1: Schematic diagram for "Quick Henry"

The audio input frequency is varied and, as the resonant frequency of the parallel combination of the unknown L and the selected C is approached, the voltage across BP4-BP5 increases sharply. At its resonant frequency the parallel L-C circuit offers its highest impedance to the incoming signal. This ac voltage across the unknown inductor is rectified by diode Dl and passed on to the sensitive dc microammeter, Ml, and the RF-blocking choke, Ll. When using an RF generator, the signal is connected to BP1 and BP2, and coupled to the resonant circuit by isolation capacitor Cl.

Once the resonant frequency of the L-C combination is known, you can determine the approximate inductance using one of three methods described later: charts, formula, or computer program.

Capacitors C2 through C5, selected by Sl, allow a broad range of L-C ratios and combinations. The OPEN position of Sl allows you to measure external resonant circuits, or to "trim" an inductor with exactly the value of capacitance needed for resonance at a desired frequency.

Resistor R1 and capacitor Cl are used to prevent the signal generators from being loaded down by the L-C circuit, which has a very low impedance when not in resonance. Diode D1 must be a germanium diode, since it conducts with a much lower voltage loss than a more common silicon signal diode. RF choke L1 is needed to prevent the meter from acting as a short circuit to the L-C circuit at high resonant frequencies.

Construction

The author's with-meter unit was built over 30 years ago with parts commonly available at that time. Figure 21-2 shows an external view of this unit. The exact enclosure and meter used then are no longer available, but are not critical. You may have an appropriate small metal or plastic cabinet and sensitive meter in your junk box, or can retrieve them from some other equipment you don't need.

Figure 21-2: External view of the author's three decade old version of "Quick Henry."

In any case, construction of the unit requires no special techniques. Wiring is not critical, but don't make the leads longer than necessary, and place Sl, C2, C3, C4, and C5 near BP4 and BP5. If using a metal cabinet, be sure that all binding posts are insulated from the box; there shouldn't be any electrical connections to the box. Figure 21-3 shows the interior view of my unit.

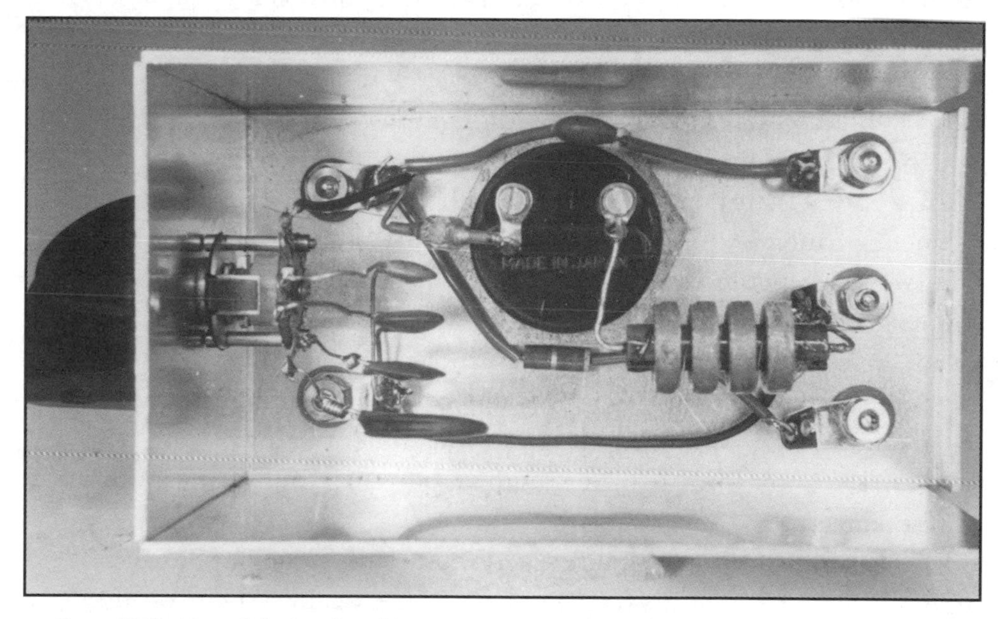
Figure 21-3: View of the interior wiring.

The five-position switch used by the author is hard to find, but Radio Shack makes a two-pole, six-position switch (see parts list) that will work fine. You just use one pole and have an extra unused position. Or you can add another "range" by using a 1 µF non-polarized capacitor in the sixth position.

If you intend to use an available external multimeter to detect the resonant condition instead of building the meter into Quick Henry, an analog meter is much preferred to a digital meter. You'll observe the analog needle move up scale to a maximum much more easily than interpreting an updating and constantly changing digital display. An oscilloscope can also be used as an external indicator.

For an external indicator, bring out two wires and clip leads for connection to the external multimeter or scope. These points are marked on the schematic as A and B. Be sure to use red (positive) and black (negative) clips to indicate proper polarity for external connections.

When using an external multimeter to sense resonance instead of the built-in meter arrangement, the multimeter should be set on its lowest DC current or voltage range, observing proper polarity.

Operation

Most inductors with many turns of wire wound on a ferrite or iron core measure over 1 mH, and can be checked using an audio generator with Quick Henry. Typical air-wound coils are in the microhenry range, so use an RF generator. Never connect both generators to Quick Henry at the same time, as all sorts of spurious signals will result.

Connect the generator "ground" lead to BP2, and the "hot" lead to either BPl (RF) or BP3 (AF). If you are using the external indicator Quick Henry, connect the clip leads to the external DC meter or oscilloscope, observing polarity. For more accurate frequency readings than your signal generator readout, you can connect a digital frequency counter directly across the signal generator leads, observing common grounds.

To start with, set Sl to the "0.00l" position. Beginning at the low frequency end of the signal generator, vary the frequency (changing generator frequency bands when necessary) until a clear meter upward deflection is observed. If using an oscilloscope, adjust the controls for high sensitivity so you can see an increase in voltage, and a fast sweep so you don't see individual cycles.

Some minor spurious responses may be seen (especially when using an RF generator and measuring in the low microhenry range), but these can be ignored. Look for a relatively high response as your proper indication of resonance.

If you don't find any significant response, try switching Sl to the next higher value (0.01), and sweep the frequencies again. When you do get a response, it will be quite definite, and might "pin" the meter (or scope display). Adjust the generator output for a comfortable peak reading.

Most audio generators have enough output to deflect a sensitive meter well beyond full scale, and RF generators will give at least 1/2 scale under most conditions. The best accuracy will be obtained with the highest value of C that gives a sharp peak meter reading, so readjust the position of Sl if necessary.

A broad peak, that is, one which is not too definite in relation to the varied frequency, is "low Q," and may be improved by using a higher value for C (setting Sl to a higher value).

Once you have found the best setting for Sl (and it really is much easier than it may sound) you can use Figure 21-4 to determine the inductance of the unknown coil if you're using an audio generator, or Figure 21-5 if you

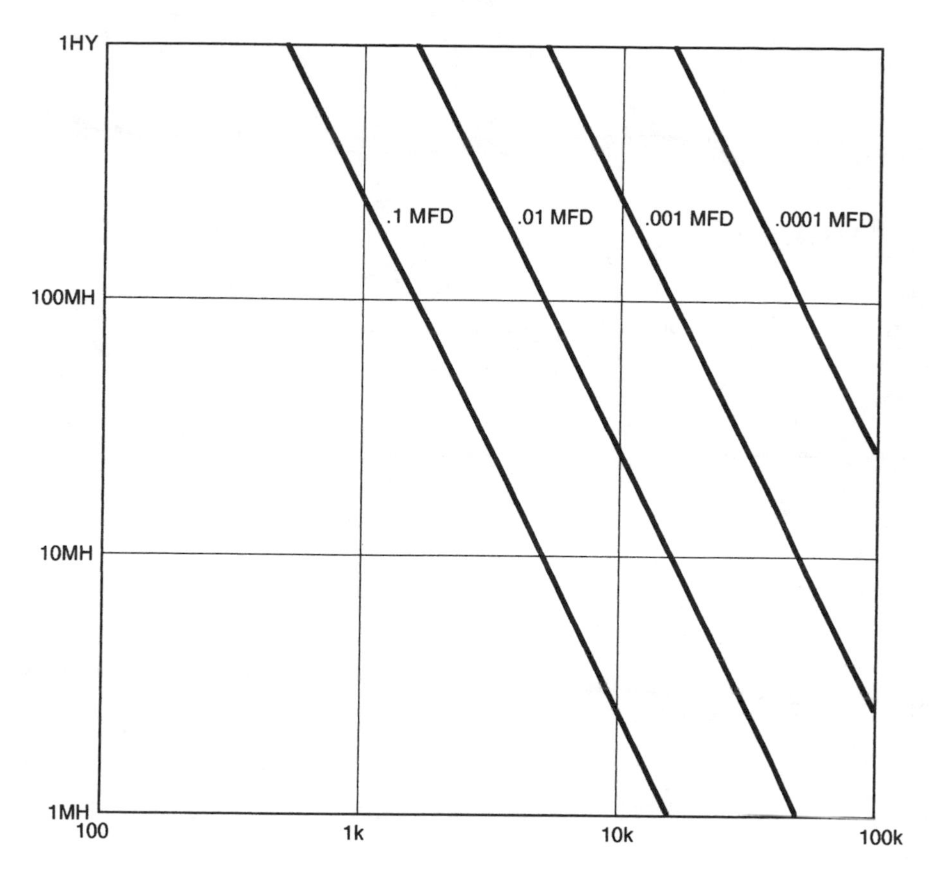

Figure 21-4: Resonant frequency versus inductance for the audio range.

are using an RF generator. Enter the horizontal scale at the resonant frequency as read on the generator dial or frequency counter and move directly upward until you intersect the line representing the value of capacitance selected by Sl. Then move straight to the left and read the value of the unknown inductance on the vertical scale.

Figures 21-4 and 21-5 have been plotted showing only the values of capacitance shown in Figure 21-1. If you desire to use any other values, either internally or connected to the binding posts externally, you

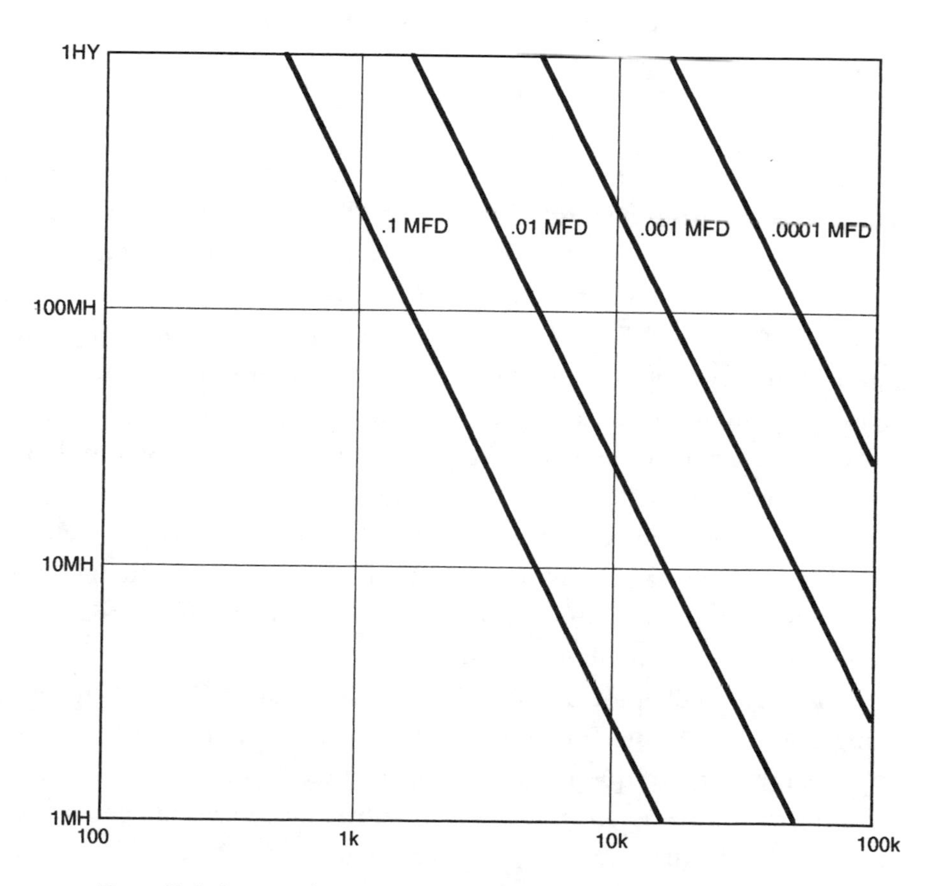

Figure 21-5: Resonant frequency versus inductance for the RF range.

can calculate the unknown inductance as follows:

$L = (1 / C(6.28F)^2)$

L = inductance in microhenries (μH)

C = capacitance in microfarads (μF)

F = resonant frequency in megahertz (MHz)

A third method to determine the inductance once the resonant frequency is known is to use the simple BASIC computer program shown in the listing. This has worked without modification with Radio Shack TRS-80 BASIC, IBM GWBASIC, QuickBASIC, and QBASIC.

```
10 REM * CALCULATE INDUCTANCE *
20 REM * F=RESONANT FREQUENCY WITH PARALLEL CAPACITOR *
30 REM * C=CAPACITOR IN PARALLEL WITH INDUCTOR *
40 REM * L=INDUCTANCE
100 INPUT"RESONANT FREQUENCY IN (1) KILOHERTZ OR (2) MEGAHERTZ";A
110 INPUT"ENTER RESONANT FREQUENCY";F
120 IF A=1 THEN F=F/1000
140 INPUT"ENTER CAPACITANCE IN MICROFARADS";C
200 L=1/(C*(6.28*F)^2)
205 IF A=1 THEN L=L/1000
210 PRINT"THE INDUCTANCE IS";L;
220 IF A=1 THEN PRINT"MILLIHENRIES":END
230 PRINT"MICROHENRIES"
```

For determining the resonant frequency of, say, an IF transformer, connect one of the windings (an ohmmeter will identify the windings by continuity) to BP4 and BP5, and set Sl to OPEN. Using an RF generator, find the frequency that peaks the meter, and read this resonant frequency right off the generator dial. With S1 in the OPEN position, a particular value of capacitor may be placed in parallel with an unknown inductor, and the resonant frequency determined in the same way.

Series resonant circuits can also be measured across BP4 and BP5 by noting a dip in the meter reading rather than an increase. When not in resonance the reactance of a series L-C circuit is high and the meter reading is high since most of the input signal is going through the meter. However, the series L-C reactance drops to close to zero at resonance, effectively almost shorting out the meter circuit.

You can actually plot the audio band pass of an R-C or L-C network by taking successive meter readings near resonance and plotting them on graph paper, with frequency along the horizontal axis, and meter reading along the vertical axis. In this case, it is convenient to set the generator output to read full scale on the meter at resonance.

Finding Q

To find the Q (relative bandpass sharpness) of a resonant circuit, you need to determine the frequencies at which the meter reading is approximately 70% of the maximum value. These are the 3dB-down power settings.

Set the peak meter reading to full scale by adjusting the generator output. Now vary the input frequency on both sides of the resonant frequency to the points where the meter reads about 0.7 of full scale. Note the frequencies where these meter readings occur and apply the following simple formula:

$Q = f_R/(f_H - f_L)$

where:

f_R = resonant frequency

f_H = high frequency 3dB down

f_L = low frequency 3dB down

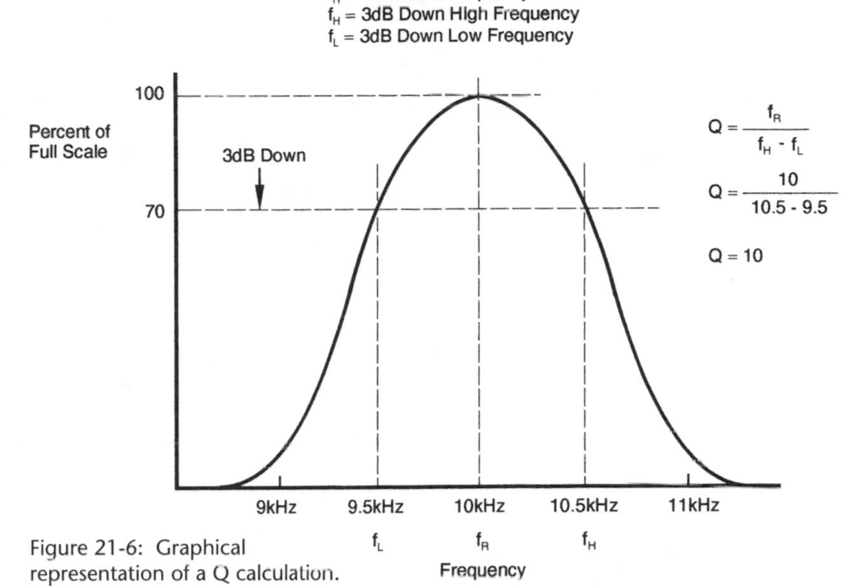

Figure 21-6: Graphical representation of a Q calculation.

Figure 21-6 shows an example of a Q calculation.

The accuracy of Quick Henry does not qualify it as a laboratory standard by any means. Numerous errors are cumulative, such as distributed capacitance, stray inductance, the external capacitance of test leads, and the calibration accuracy of the signal generators used. However, for the home experimenter or the small repair shop or lab, which do not usually need extreme accuracy, Quick Henry will satisfy a need to measure inductance, resonance, and Q quickly and inexpensively.

The Radio Shack and Mouser Electronics parts numbers indicated above, as well as the availability of those parts, may have changed since publication of this book.

PARTS LIST

R1:	4.7 kilohm 1/4 watt resistor (not critical value)
C1:	20 pF ceramic disc capacitor (not critical value)
C2:	0.1 µF ceramic disc capacitor (Radio Shack #272-135)
C3:	0.01 µF ceramic disc capacitor (Radio Shack #272-131)
C4:	0.001 µF ceramic disc capacitor (Radio Shack #272-126)
C5:	0.0001 µF (100pF) ceramic disc capacitor (Radio Shack #272-123)
D1:	IN34A germanium diode (Radio Shack 276-1123)
S1:	Five position switch (see text; Radio Shack #275-1386)
L1:	2.5 mH RF choke (Mouser Electronics #434-2250)
BP1-BP5:	Insulated binding posts (Radio Shack #274-662)
M1:	0-100 microampere meter (see text)
Case:	See text
Selector Knob:	To fit S1 shaft (Radio Shack #274-424)

CHAPTER 22

A Deluxe Timebase

This "Deluxe Timebase" features five separate square-wave outputs: 100 Hz, 1 kHz, 10 kHz, 100 kHz, and 1 MHz. It can be built inside a plastic case small enough to fit in your pocket. It can be built from mostly common parts or with a parts kit described at the end of this chapter.

If you have ever wondered about the accuracy of the dials and digital readouts on your frequency-related test equipment, this deluxe timebase circuit can provide you with a simple and inexpensive means of calibration. Although it is not up to National Bureau of Standards accuracy, it is certainly close enough for most experimentation and repairs.

You can use the timebase directly to closely calibrate digital readouts of AM and FM radio receivers up to 30 MHz and probably higher. It can be used as a signal injector to trace signals in audio and radio frequency amplifiers. Used with an oscilloscope and Lissajous figures, as described later in this chapter, you can calibrate AM audio and RF signal generators. You'll probably find other uses as well.

Circuit Description

Figure 22-1 shows the timebase schematic. The heart of the circuit is the oscillator, X1. This integrated circuit crystal-controlled oscillator, when properly powered through current-limiting resistor R1 to pin 14, generates a square-wave output at one million cycles per second (1 MHz), plus or minus 50 Hz, at pin 8. Since the internal oscillator in X1 is crystal controlled, the output signal is very stable and precise.

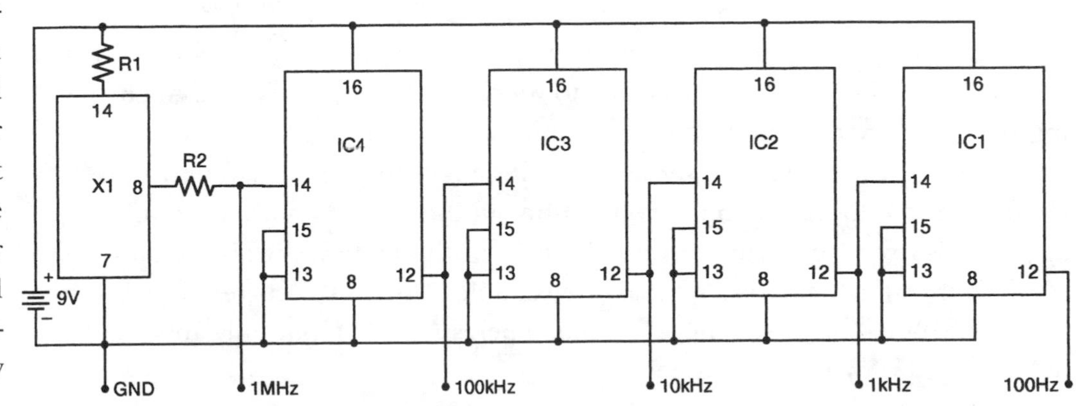

Figure 22-1: Schematic diagram of the deluxe timebase.

The 1 MHz output signal of X1 is connected through current-limiting resistor R2 to pin 14, the clock input of IC4, a CMOS 4017B divide-by-10 counter. The counter divides the input signal by 10, producing an output signal on pin 12 of 100,000 cycles per second (100 kHz).

This signal is also fed directly to the clock input of IC3, another 4017B decade counter, resulting in a 10 kHz output. A similar divide-by-10 process is produced by IC2 and IC1, resulting in outputs of 1 kHz and 100 Hz respectively.

Construction

You will have to look long and far to find a project this useful that is so simple to build. While a printed circuit board is not required, the layout for the one supplied with the parts kit is shown in Figure 22-2. The parts layout, which could just as easily be used with common perforated board having holes placed 0.1-inch apart, is shown in Figure 22-3.

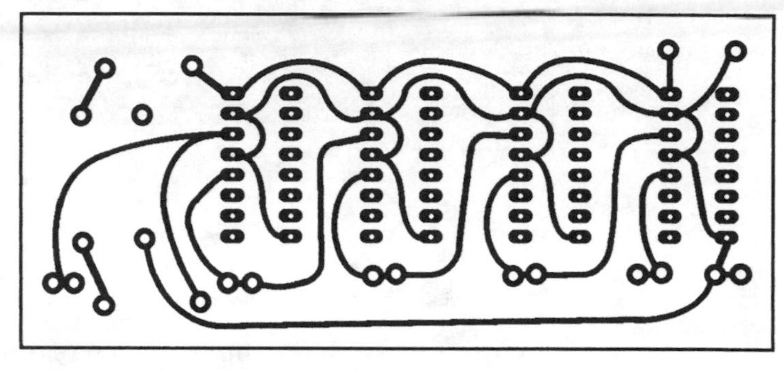

Figure 22-2: Circuit board layout for the timebase.

The integrated circuit crystal oscillator, X1, is currently available from Mouser Electronics (see parts list) for $4.88 as this is written. It is also included in the parts kit along with all other required parts and an etched and drilled printed circuit board.

While 16-pin sockets for IC1–IC4 come in the kit, they are not actually required; you will only be soldering to six pins of the 16 pins on each chip. However, if one of the ICs goes bad, you'll be glad you used sockets!

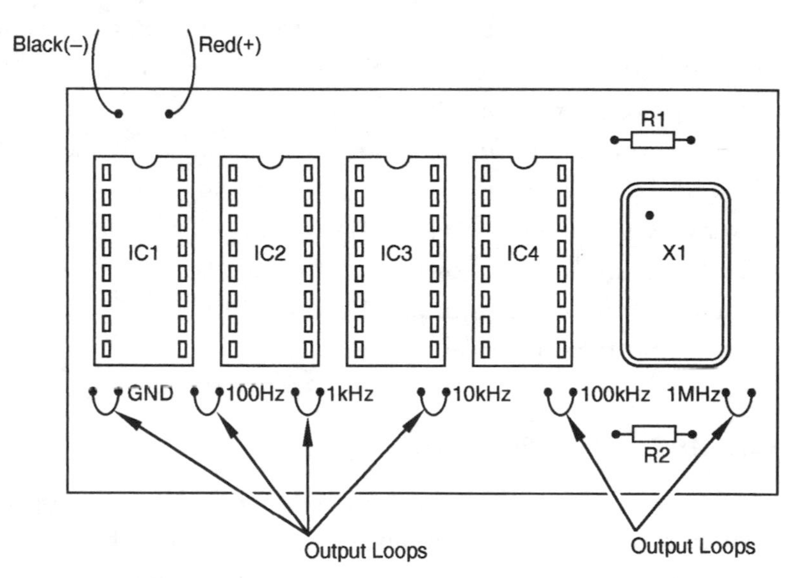

Figure 22-3: Parts layout and placement on the circuit board.

With only two resistors, the crystal oscillator, and four integrated circuits, the assembly and wiring of the circuit board is relatively simple. Be sure the ICs and crystal oscillator are oriented as shown in Figure 22-3, and that the wires going to your battery are not reversed. Applying reverse voltage to the ICs or crystal oscillator could ruin them instantly.

Although the schematic shows a 9 volt battery for power, I found the circuit worked down to about 3 volts, although the output square wave voltage was also reduced. The current drain from the batteries is also greatly reduced if you use less than 9 volts. For example, I found the current drawn from a 9 volt battery was almost 15 milliamperes (15 mA), but only about 2 mA was used with a 3 volt source.

Deluxe Packaging

While the project can be used in its bare-bones configuration, I decided to make it more practical to carry around. Any plastic box can be used, but I used a "Fuji slide box" as described in other chapters. It is the perfect size for this project (4.5 by 2 by 1 inches), made from thin easily-cut plastic, and holds the circuit board and 9 volt battery perfectly.

I decided to use a two-pole six-position rotary switch to select the output frequency fed to two binding posts. The switch also is used to power the circuit, with an LED to show the unit is ON. Figure 22-4 shows a wiring diagram of the added circuitry. The switch, binding posts, resistor, and LED were mounted on the box cover, with the battery and circuit board inside the box, as shown in the photos. If the switch comes with a long shaft, you'll want to cut it shorter with a hacksaw, leaving it long enough to use with a set-screw type pointer knob.

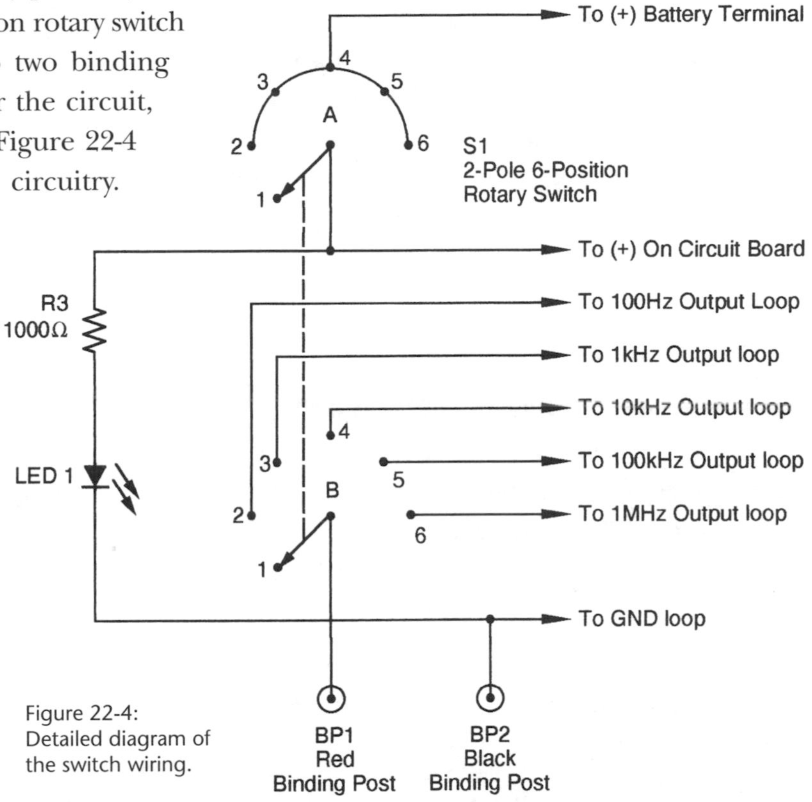

Figure 22-4: Detailed diagram of the switch wiring.

You must take care that you wire things properly to the switch. Use an ohmmeter to determine which switch terminals are in contact at each switch position, and make a pictorial diagram of the switch showing these connections. Referring to Figure 22-4, run wires from the circuit board to the proper switch terminals. I found flexible "rainbow" ribbon cable (adjoining wires are different colors) easy to use between the flip-open cover and the circuit board.

When connecting the LED and its current-limiting resistor between the switch and the negative binding post, be sure the LED cathode side (usually a flat spot at the base of the LED) is connected to the binding post. Solder lugs on each binding post simplify assembly.

Figure 22-5 shows the completed circuit inside a Fuji slide box. Note the additional wiring around the switch; this is why I stressed the need to take care when wiring the switch.

Figure 22-5: Inside of the "Fuji slide box" showing the switch wiring.

Testing

Once the unit is assembled, turn the switch to its furthest counter-clockwise position and tighten a set-screw type pointer knob to the round switch shaft so that the knob points to the left. This is the OFF position. Connect a 9 volt battery to the battery snap. When you twist the knob clockwise to the first position, the LED should light, and a 100 Hz square-wave should appear at the binding posts. You can use a typical 8 ohm ear-phone to monitor the output in this position. Alternately (and better) is to use an oscilloscope.

As you turn the knob clockwise, each detent raises the output frequency tenfold, such as 1 kHz, 10 kHz, 100 kHz and finally 1 MHz. Your earphone probably will not have sufficient response at 10 kHz or above, but your scope should easily show these signals.

Troubleshooting

If you get no output at all, make voltage checks. Be sure each IC is oriented as shown in Figure 22-3. If the outputs seem reversed in frequency—going down as you turn the switch clockwise—you have mis-wired to the switch. If so, just switch wires at the output loops on the circuit board. If the LED does not light, you may have the cathode wired to the switch instead of the negative terminal.

Using the Deluxe Timebase

Okay, you have a properly working timebase. What are you going to do with it? To check the accuracy of the dials and readouts on radio receivers, use the timebase with a simple output coil as a "marker generator."

Suppose, for example, you want to check (or set) the dial or digital readout of an AM radio. Make an output loop from some hookup wire. The number of turns and dimension are not critical. I used six turns of bare wire around a yellow pencil, leaving about two inches of bare wire at each end, then spaced the turns so the length of the coil was about 0.5 inches. I connected the ends of the coil to the timebase binding posts, so the coil now became the output "antenna" to our timebase "transmitter." Due to the huge harmonic content of the square-wave output of the timebase, it radiates high multiples of its frequency setting.

By turning on the timebase and holding the coil near a radio's antenna or innards, you'll be able to identify marker frequencies by either drastic quieting of the noise level, or whistling (heterodynes) caused by a received signal close to the timebase frequency. You'll find that moving the timebase coil closer and further from the radio will help you clarify if the quieting or whistling you hear is resulting from the timebase marker.

Although this technique won't tell you the actual frequency, it will identify spaced markers based on your setting of the Timebase switch. Set the

Timebase to 100 kHz for the AM broadcast band, 1 MHz for shortwave bands or amateur "ham" receivers.

Marking FM broadcast frequencies works the same way, though not quite as well, since you are into the 88 MHz to 108 MHz band, and therefore using the 88th to 108th harmonic of the 1 MHz timebase oscillator.

By connecting the output binding posts directly to audio amplifier circuits with clip leads, and setting the timebase switch to 100 Hz or 1 kHz, you can trace audio signals from the speaker back through the amplifier. For radio frequency amplifier applications, set the timebase switch to 100 kHz or 1 MHz and use an oscilloscope to display the square-wave signal through the components.

Used with an oscilloscope, you can use the timebase to calibrate audio frequency (AF) and radio frequency (RF) generators. The connections are shown in Figure 22-6 to produce modified Lissajous figures.

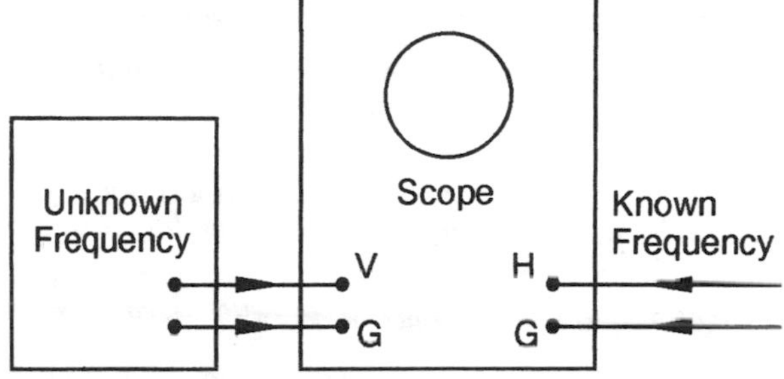

Figure 22-6: Oscilloscope connections to calibrate AF and RF signal generators.

Lissajous (Liz-a-jew) figures? These are typically used with sine waves to determine an unknown frequency. A known sine wave frequency is applied to a scope horizontal input, and an unknown sine wave frequency is connected to the scope vertical input. Once the vertical and horizontal voltages are adjusted to produce a centered scope pattern of about equal vertical and horizontal size, either input frequency is varied until a circle appears on the screen. The circle may rotate, or you may get other patterns, as shown in Figure 22-7, for multiples of the vertical or horizontal frequencies. For a full discussion of Lissajous figures, refer to a book on oscilloscopes. The purpose here is to describe their use, not their theory.

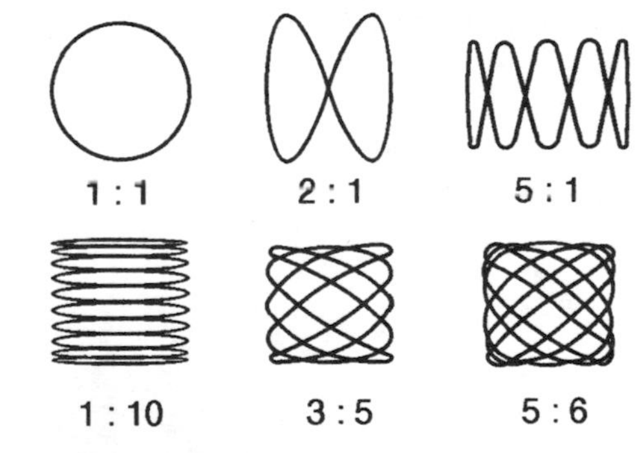

Figure 22-7: Different Lissajous figure patterns.

The known frequency (in this case, the timebase) is connected to the scope horizontal input, and the unknown variable frequency (such as an audio generator) is connected to the vertical input. Very important: the scope horizontal sweep frequency control should be set to HORIZONTAL INPUT.

Since the timebase has a square wave output, not a sine wave, you won't get the nice patterns shown in Figure 22-7. You'll see two vertical parallel lines. When the unknown frequency is varied and approaches the timebase frequency, the parallel vertical lines start oscillating slightly up and down, first quickly, then slowly, then stop. That's a match! On better scopes you'll see horizontal lines as well, and a perfect frequency match is a square with a little twist.

Using this technique and your timebase, you'll be able to calibrate the dial settings of RF generators as well, but only up to the frequency limitation of the scope. For older scopes, it might be just a few megahertz before the gain drops so much that you get too small a scope image—or just a dot!

With a little imagination, and if you like to experiment, you'll be surprised at how many radios will pick up the timebase signal, how effectively you can trace signals through an amplifier, and the variety of Lissajous patterns you can display on your oscilloscope.

Figure 22-8 shows an external view of a timebase that I packaged inside a Fuji slide box. Some transfer labels give it a professional appearance.

Figure 22-8: External view of a completed unit.

A parts kit, which includes all the ICs, sockets, R1, R2, X1, battery snap, and a drilled and etched printed circuit board (but not the additional parts shown for this project) is available from Electronic Goldmine, P.O. Box 5408, Scottsdale, AZ 85261; telephone (602) 451-7454. Their catalog number is #C6717. Parts numbers and availability for this and the Radio Shack components indicated above may have changed since publication of this book.

PARTS LIST

BP1, BP2:	Red and black binding posts (Radio Shack 274-662)
IC1, IC2, IC3, IC4:	CMOS 4017B integrated circuit
LED1:	Red LED, standard size
R1, R2:	100 Ω 1/8 watt 10% carbon resistors
R3:	1 kilohm 1/8watt 10% carbon resistor
X1:	1 MHz crystal-controlled oscillator IC (Mouser 520-TCF100; see text)
SW1:	Rotary switch, two-pole six-position (Radio Shack 275-1386)
Miscellaneous:	Etched and drilled printed circuit board, Fuji slide box, four 16-pin sockets, battery snap, knob, 6-inch bare wire.

Part Substitution

If you build electronic devices from this book or magazine construction articles, or if you repair electronic devices for yourself or others, you have run into the problem of part substitution. While this can be a small problem with common discrete components, it becomes a big problem to find equivalent parts for many transistors, diodes, and integrated circuits. The information that follows should be very useful.

Repairing radios 50 years ago was a relatively simple task. A burned out resistor, a leaky capacitor, a defective vacuum tube, a poor solder connection, or any of a number of other relatively simple problems could be found quickly by observation or with a signal tracer (which was nothing more than a common audio amplifier) or a multimeter.

When television came along, things got more complex. Much higher voltages and more sophisticated circuits made an oscilloscope almost a necessity for troubleshooting. But, once the defective part or parts were determined, the actual repair still pretty much depended on a limited number of discrete components—tubes, coils, resistors, capacitors, etc.

Audio and video tape recorders added greatly to the complexity of repairs. They had more special circuitry, and many interacting mechanical parts with their associated wheels, gears, springs, belts, and sensors. In the last 20 years computers, camcorders, VCRs, and all manner of electro-mechanical devices have come on the scene.

Three major changes in electronics in the last 50 years have resulted in greatly increasing servicing complexity: printed circuitry, semiconductors, and miniaturization (including surface-mounted components).

The small size and enormous capabilities of semiconductors, together with the techniques developed in producing and stuffing multi-sided printed circuit boards, have resulted in a huge decrease in size and increase in component density. Compact packaging has made it very difficult to access and test many parts at the discrete level. Much "repair" these days consists of simply replacing complete modules or circuit boards rather than tracing the problem to an individual component and replacing it.

Discrete Components

However, there are still many electronic equipment malfunctions that can be repaired with simple discrete components. Once it is determined to be the culprit, a burned resistor, or a shorted or open capacitor, can usually be unsoldered and replaced, even if mounted on a PC board. Defining the requirements to replace the failed part can usually be determined by the size, shape, color code, or other markings on the failed part. The device schematic (if you have one) will tell you the proper value needed.

Equivalent replacement parts for resistors, capacitors, relays, inductors, and other common electronic parts are readily available from Radio Shack stores or regular electronic part distributors.

When replacing a resistor, the resistance value and the wattage rating are the main concern, although tolerance is sometimes important. When substituting potentiometers not only is the maximum resistance important, but the "taper" (audio or linear resistance change with rotation), single-turn or multi-turn, size, type of mounting, and wattage are primary concerns.

Substitution of capacitors is tricky, since there are so many types—disc, electrolytic, tantulum, mylar, variable, trimmer, and many more—and their characteristics are different. Of course, the value of capacitance and tolerance are important, but be sure the voltage rating is adequate for the circuit.

Relays and inductors usually have less flexibility in replacement, since the physical size may be limiting. For relays, the coil resistance or voltage, AC or DC, contact arrangement and contact voltage rating are all very important. Inductors vary in size, inductance value, and voltage rating, with tolerance usually least important.

It is when dealing with semiconductors—diodes, resistors, integrated circuits and other "solid-state" components—that you frequently find yourself facing many unknowns.

Semiconductors are produced by many, many manufacturers all over the world, most with numbering systems unique to the manufacturer. Three major manufacturers I'll highlight here specialize in providing semiconductor part substitution; they have huge data books and computer software that allow you to find equivalent parts.

NTE/ECG/SK Substitutions

Literally tens of thousands of semiconductors with various characteristics are offered by dozens of large manufacturers around the world. How can you determine what transistor (for example) to use to replace the one that has failed?

Well, it turns out that vast numbers of semiconductors are so similar that several manufacturers each offer a reduced number of "equivalents" that cover a broad range of individual part requirements. In particular, I'll discuss the semiconductor replacements offered by NTE Electronics, Philips

ECG, and the Thomson Consumer Electronics SK-Series. I'll cover their manuals and computer search software.

See the sidebar for addresses and phone numbers to contact for ordering information.

General Rules for Substitution

Whenever you need to replace a semiconductor, you can consider it safe to replace it with a part having the identical part number and manufacturer. But this is often not possible. For one thing, it might be a foreign semiconductor not readily available from your normal parts source or, it might be an old part no longer manufactured.

This is when you are forced to look for a functional substitute, and this is where things can get difficult. Semiconductors have different critical characteristics, such as maximum voltage, current, power, and gain. Specific circuits can have certain critical requirements, such as frequency response, trigger current, linearity, power limit, gain, etc. What you need depends on your circuit application. Here are some simple rules to consider, as they apply in each case, when looking for a substitute semiconductor:

- Use equal or greater breakdown voltage.
- Use equal or greater operating current.
- Use equal or greater power dissipation.
- Use equal or greater gain bandwidth product.
- Use equal or lower switching time.
- Use equal or lower trigger current.
- Use equal or lower reverse current.

Manuals and Software

NTE (NTE Electronics, Inc.), ECG (Philips ECG), and SK (Thomson Consumer Electronics) parts are described in detail in thick manuals available for low cost from each of these manufacturers. In addition, each company offers computer disks with their database of available parts and their equivalents.

You'll probably need both the manual and the disk for each company whose parts you decide to use. While the disks provide very fast access to part number equivalents and limited specifications, you'll need the manuals for detailed specifications and pinouts.

A typical "NTE Semiconductors" catalog covers 3,500 NTE semiconductors cross-referenced to over 250,000 U.S., Japanese, and European part numbers, including technical information. The computer database, "QUICKCross," is available on 5.25-inch or 3.5-inch diskettes for MS-DOS, and 3.5-inch diskettes for Windows. Contact NTE for current prices (around $10 at the time this was written). NTE also offers relays, resistors, capacitors, and flyback transformers. Their Internet address is http://www.nteinc.com.

A typical "ECG Semiconductors Master Replacement Guide" covers nearly 4000 ECG products, and features 276,000 cross-references. The computer database, "INSTANT CROSS Version 2.0" (MS-DOS) consists of three IBM PC 720K diskettes with a suggested retail price (as of 1997) of $11. ECG also manufactures many other electronic products, including capacitors, circuit breakers, test equipment, connectors, and surge suppressors.

The "SK Replacement Cross-Reference Directory" offers 3,600 replacements for over 235,000 parts, including NTE and ECG parts. The computer database, "SK Cross Version 3" (MS-DOS), consists of two 1.44MB diskettes with a suggested retail price in 1997 of $5.

Since each of the manuals is intended to provide the same information about their line of semiconductors and the equivalents, they approach the task in a similar manner with slight variations. The diskettes that contain their databases come in 5.25-inch or 3.5-inch sizes with different densities and different prices. Contact each company for the details.

Using the NTE Manual and Database

"NTE Semiconductors – 6th Edition" has two tabbed parts: Technical Guide and Semiconductor Cross-Reference. The Technical Guide portion begins with an alphabetical and numerical product index section, then devotes several pages to replacement considerations, including handling and mounting techniques. This is followed by sections of data on various types of semiconductors: transistors, diodes/rectifiers, high voltage devices, industrial rectifiers/bridges, SCRs/TRIACs, powerblock/hybrid modules, special devices, transient suppressors, IC protectors/thermal cutoffs, optoelectronic devices, voltage regulators, linear integrated circuits, digital integrated circuits, microprocessors/memories, etc. This is followed by sections on flameproof resistors, various types of capacitors, sockets, and heat sinks. Each of these sections includes dimensioned outline drawings and pinouts of each part, making this a valuable reference book.

However, it is the alpha-numeric Semiconductor Cross-Reference portion (over 300 pages in tiny print with seven columns per page!) that leads you to the NTE equivalent part numbers. You look up the part number you have and the NTE replacement part number is shown. Next you go to the Technical Guide numerical index to find the page number for a detailed description of that part and its operating characteristics, and the diagram number for the part dimensions and pinouts.

The "QUICKCross" database comes with installation and user instructions. I used Version 3.1 for MS-DOS, but now Version 4 for DOS will be available, as well as a Windows version. I found the DOS installation of four 720K diskettes to the hard drive took only about five minutes, and occupied a little over 3 megabytes of hard disk space. You'll need 512K of free RAM and DOS 3.3 or better. Technical assistance, if necessary, is a phone call away. I had no questions, since the program ran very quickly and easily. The program runs in color, with no default to monochrome.

Using QUICKCross could hardly be easier. After a couple of initial information screens, you can select what you want from a Main Menu. The database on these disks includes not only semiconductors, resistors, and capacitors, but also has relay and flyback transformer cross-references. (Technical details and drawings on these last two are available in separate NTE manuals.) FAX support is also described.

If, for example, you want to know the NTE substitute part for a 2SA604 transistor, you can select "Semiconductor Cross-Reference" from the Main Menu, and you'll get a screen that will allow you to type in three or more characters of the desired search number. Almost instantly the screen will display NTE129 as the replacement, and will also indicate this is a PNP silicon transistor used for AF output and packaged in a TO-39 case. For further details, you would refer to the NTE Semiconductors manual. If there should be other replacements, another screen pops up with a list showing them.

The five cross-reference and selection choices on the Main Menu all work the same way: simple and fast. But you'll need the manuals for details-specifications, outline, pinouts—if you want those details.

The Information Section offers a short history of NTE Electronics, tells you how to get technical literature "for the asking," locates the NTE sales offices in your state or internationally, allows you to print out an Information Request Form, and even provides instructions and encouragement for you to COPY these disks and distribute them!

QUICKFACTS Customer Fax Support gives you the phone numbers and document numbers to "fax-back" the index to each of their product lines. Navigation around the screens (which are all text-mode in the DOS version) is accomplished easily, and the Escape key gets you back if you head down the wrong path. Also you can print most screens very easily. In general, I liked this software better than the others tested.

Using the ECG Manual and Database

The "ECG Semiconductors Master Replacement Guide – 16th Edition" has a 368 page technical section, a 310 page cross-reference section, and 11 additional pages describing various other Philips products. The technical section begins with an alphabetical and numerical product index, then devotes six pages to replacement procedures, including handling, mounting, identifying and testing techniques. This is followed by sections of data on various types of semiconductors: transistors (including surface mount), high voltage devices, diodes/rectifiers, SCRs/TRIACs, special purpose devices, transient suppressors, crystals, opto devices, thermal cutoffs/IC protectors, industrial power modules, linear ICs, digital ICs, interface/MPU ICs, and accessories/heat sinks. Each of these sections includes dimensioned outline drawings and pinouts of each part.

The over 300 page small-print alpha-numeric cross-reference portion (seven columns per page) allows you to find the ECG equivalent part numbers. You must go to the technical section for a detailed description of that part's operating characteristics, diagram number, and pinouts.

ECG's "INSTANT CROSS" database comes on three 3.5-inch 720K diskettes with a small card describing the installation procedure. It takes about ten minutes to install the diskettes onto your hard drive, and about 3.5MB of hard disk space is used.

The program is extremely easy to use, and can be customized with your name and address, as well as many color combinations, or displayed in black and white. The screens are all DOS text with borders. The Escape key quickly backs through the choices if you find yourself where you don't want to be.

After some introductory screens you are at the Main Menu where you can select finding the ECG replacement number, finding an ECG type description, printing the complete list of parts with abbreviated descriptions (about 250K!), or go to a Utility Menu. Finding a replacement number is as simple as typing in three or more characters and hitting the Enter key; the replacement number or numbers pop up almost instantly. At the bottom of the screen is a short description of the highlighted item. You can print the description of the highlighted number by just hitting the F4 key. For further details (specifications, pinout, dimensioned outline drawing) you must refer to the ECG Semiconductors Master Replacement Guide.

The Utility Menu choice allows you to select either color or monochrome monitor, making the program compatible with older monochrome desktops and laptops. If you have a color display, you can choose from eight foreground, background, highlight text, and highlight background colors. Another nice feature is that you can minimize the startup time each time you run the program by shortening the display time of the initial screens.

Using the SK Manual and Database

The "SK Replacement Cross-Reference Directory" (last edition, as of 1997, is dated 1992) starts with a subject index followed by an alphabetical and a product index, then a competitive cross-reference to Philips ECG and NTE products. This is followed by technical sections on transistors, rectifiers, thyristors, optoelectronics, integrated circuits, telecommunication circuits, CMOS/MOS/NMOS/PMOS, high speed CMOS, DTL and TTL logic, microprocessors and memories, flameproof resistors, hardware, sockets, heat sinks, accessories, and voltage regulators. An exact semiconductor replacement guide precedes the main cross-reference of equivalent replacement parts.

Thompson's "SK Cross" program comes on either 5.25-inch 360K or 3.25-inch 1.44MB diskettes. I used the latter, and had no trouble installing the program on my hard drive in less than ten minutes, although the program expanded to almost 12 megabytes of hard drive space during installation. You must have at least 310K of free conventional memory, a very modest requirement.

SK Cross was incredibly fast and easy to use. There is no main menu. You just type in one to ten of the first characters of the desired part number and the screen instantly displays a sequential list of 14 industry part numbers. The list starts with the characters you entered, or the next industry part number in the database. You can scroll up or down the list to other nearby part numbers.

On each screen line you see the industry part number, the SK equivalent part number, and a short description. Highlight the line you want, hit Enter, and the screen displays additional information on that SK part. However, for detailed specifications, outline drawings and pinouts, you must refer to the cross-reference directory.

If you start the program by typing "TCE" (no doubt for "Thomson Consumer Electronics"), the program displays in color. If you start with "TCE M" the program displays in monochrome, making it usable with older desktop and laptop computers. There is no special provision to print out results, but you can simply press the "Print Screen" key on your keyboard to print any screen in text and borderline characters. The screen reverse highlighting is ignored.

Do You Need the Disks?

Although the software is handy for finding numbers fast, you'll still may need to refer to the thick manuals for specifications, pinouts, and outline drawings to choose a replacement part for your particular use. If you have a frequent need, such as working at a parts house where you supply substitute parts to local electronic technicians, the software will get you a fast replacement part number, but you may still have to refer to the company's cross-reference directory for the physical package and technical limitations of the replacement part.

The manuals are extremely useful references, and NTE, ECG, and SK parts all will do the job, properly chosen. When it gets down to the bottom line, the specific electronic application and your judgment make the difference between a proper or marginal replacement.

Call the following manufacturers regarding current prices and ordering information for their cross-reference manuals and disks:

NTE Electronics, Inc.
44 Farrand Street
Bloomfield, NJ 07003
800-631-1250 or 800-683-6837

Philips ECG
112 Polk St.
P.O.Box 309
Greeneville, TN 37744-0309
Customer Service: 800-233-8767
(Will refer you to your local dealer.)

Thomson Consumer Electronics
2000 Clements Bridge Road
Deptford, NJ 08096-2088
Customer Service: 800-635-2474
(Will refer you to your local dealer.)

Kit and Part Sources

Various kit manufacturers have been listed throughout this book. Here I conveniently list each of them, with additional information, such as shipping charges, Internet address, etc. at the time this is written—all of which may change in time.

For those of you that prefer to build "from scratch" but need some parts you don't have, I'll describe several part sources.

All prices and information are at the time of this writing. Catalogs are available free by calling the listed order numbers.

Kit Sources (Listed alphabetically)

Cal West Supply, Inc.
31320 Via Colinas #105
Westlake Village, CA 91362
Orders/Info: (800) 892-8000, (818)889-2209
Fax: (818) 706-0825
Internet: http://www.hallbar.com
E-mail: hallbar@hallbar.com

This company offers 44 HALLBAR electronic kits, most with printed circuit boards and cases, and electronic components. Components may be ordered separately. Minimum order (before shipping or tax) is $10. Shipping and handling $4.50 USA; Alaska, Hawaii and Canada orders must include 20% for shipping (minimum $6) with excess over actual shipping refunded. California residents add 8.25% sales tax.

C & S Sales, Inc.
150 W. Carpenter Ave.
Wheeling, IL 60090
Orders: (800) 292-7711
Inquiries: (708)541-0710 Fax: (708)541-9904
Internet: http://www.elenco.com/cs_sales/
E-mail: cs_sales@elenco.com

They offer a broad range of test equipment from various well-known manufacturers, as well as robot kits, meter kits, radio theory kits, and 44 educational kits. Minimum order is $25, with a 5% ($5 minimum) for UPS ground shipping and handling charge for USA orders, or 10% ($8 minimum) for Canada. Alaska, Hawaii, Puerto Rico and foreign inquire. Illinois residents add 8% sales tax. 15-day money-back guarantee. (NOTE: C & S Sales sells Elenco products at a discount, NOT presently shown on the website. Call or e-mail for C & S price.)

Centerpointe Electronics, Inc.
5241 Lincoln Avenue, Unit A6
Cypress, CA 90360
Orders/Info: (800) 272-2737
Fax: (800) 493-7862
Internet: http://www.shopsite.com/electronics
E-mail: ctrpoint@ix.netcom.net

They offer 76 CTI, Cana-Kit, and CEKIT kits, plus parts for their 30-IN-ONE LAB and DIGITAL LAB. They also offer electronic equipment, and parts, including integrated circuits and transistors. No minimum order noted. Shipping 5% or $5 minimum. California residents add 7.75% sales tax. Free full-color wall chart of all kits available by calling (800) 422-1100.

Electronic Goldmine
P.O. Box 5408
Scottsdale, AZ 85261
Orders: (800)445-0697
Inquiries: (602)451-7454
Fax: (602)661-8259
Internet:http://www.goldmine-elec.com
E-mail:goldmine-elec@netwrx.net

Their 1997 catalog shows 108 electronic project kits, plus innumerable standard and special parts that would make an experimenter think he was in heaven! Minimum order is $10 plus shipping. Shipping and handling is $5 minimum for up to three pounds in USA. Call for additional shipping information. Arizona residents add 6.95% sales tax.

Electronic Rainbow, Inc.
6227 Coffman Road
Indianapolis, IN 46268
Orders: (888) 291-7262
Tech/Info: (317) 291-7262
Fax Orders: (317) 291-7269
Internet: http://www.rainbowkits.com
E-mail:eri@iquest.com

They offer over 50 electronic kits (some available assembled), as well as electronic tools and parts. The Kit Book, which contains schematic, parts list and printed circuit board layouts for many Rainbow kits, sells for $14.95 (or $9.95 with any kit purchase). No minimum order. UPS ground shipping is $5 for the first pound, $1 for each additional pound for U.S. customers; call for other shipping. Indiana residents add 5% sales tax.

Ramsey Electronics, Inc.
793 Canning Parkway Victor, NY 14564
Orders: (800)446-2295
Technical: (716)924-4560
Fax: (716)924-4555
Internet: http://www.ramseyelectronics.com

Ramsey offers over 80 kits, many with various versions. Some are very elaborate, and some are available wired and tested. Of most interest to readers of this book looking for simple kits would be Ramsey's Mini-Kits. The catalog includes many amateur radio kits, personal radio broadcasters, video cameras and transmitters, and many hobby kits, as well as tools and test equipment. Minimum order is $10, with an additional $3 small-order charge for orders less than $19.95. USA shipping, handling and insurance charge is $5.95. Call for Canada and overseas shipping. New York State residents add sales tax.

Whiterook Products Company
309 S. Brookshire Ave.
Ventura, CA 93003
Orders/Info/Fax: (805) 339-0702
Internet: http://www.west.net/~wpc

Whiterook offers many unique amateur radio items and unusual 6-digit or binary digital clocks in kit form as well as assembled. Their Internet site is one of the nicest I've seen, with colorful quick-loading line-art and text describing each product, and many interesting links. No minimum order. Shipping and handling per order is $6.50 within Continental USA, $10 to Canada; no foreign orders. California residents add 7.25% sales tax.

Part Sources (Listed alphabetically)

Digi-Key Corporation
701 Brooks Avenue South
Thief River Falls, MN 56701-0677
Orders: (800) 344-4539
Technical: (218) 681-6674
Fax: (218) 681-3380
Internet: http://www.digikey.com

Their large catalog covers thousands of parts in 15 major categories from 183 different manufacturers. They offer same-day shipment with no minimum order, but a $5 handling charge is added for orders under $25. No shipping charge to USA or Canada (with some exceptions) on prepaid orders. Foreign, Canada, Mexico orders add $5. Minnesota residents add 6.5% sales tax.

Mouser Electronics
National Circulation Center
2401 Highway 287
North Mansfield, TX 76063-4827
Sales/Service: (800) 346-6873 (automatically routed
to the closest of their three locations):
Santee, CA (619) 449-2222, Fax (619) 449-6041;
Mansfield, TX (817) 483-4422, Fax (817) 483-0931;
Randolph, NJ (201) 328-3322, Fax (201)328-7120
Internet: http://www.mouser.com
E-mail orders: sales@mouser.com
E-mail technical and non-catalog items: tech@mouser.com

Mouser stocks thousands of parts in 11 major categories from 139 different manufacturers. Their service is extraordinary. I'm about 150 miles from their nearest location, yet although I called one Thursday afternoon about 4 p.m., my order arrived the next day at 10 a.m.! Also, for those with computers, a CD-ROM with the complete contents of their catalog is available free on request. No minimum order or handling charge to U.S, Canada or Mexico; only actual shipping charge. Same day shipment. Call regarding foreign orders. California, Texas, and New Jersey residents add sales tax.

Newark Electronics

81 branch offices in USA, plus foreign phone: (800) 463-9275
(Automatically routed to closest office.)

Internet: http://www.newark.com

Their huge catalog shows 125,000 products in 28 categories from 300 manufacturers. If you can't find the part you want anywhere else, you can probably find it in this enormous catalog. It includes a special manufacturer part number index that takes 22-pages in small-print. A free CD-ROM of the complete catalog is available on request. $25 minimum order. Same-day shipping at actual shipping cost from closest warehouse. Add state and local sales tax.

Radio Shack

6600 stores

Phone order or nearest store: (800) 843-7422

Internet: http://www.radioshack.com

E-mail product support: support@tandy.com

The latest Radio Shack catalog is available at any of their 6600 stores nationwide. It offers thousands of products and parts you can order by phone or find at your local Radio Shack store. Plus, over 100,000 unique and specialty items and accessories are available through Radio Shack Unlimited. No minimum order. Shipping and handling charge is based on amount of purchase, with a $4 minimum. Add state and local sales tax.

TechAmerica

P.O. Box 1981

Fort Worth, TX 76101-1981

Orders: (800) 877-0072

24-Hr. Fax: (800) 813-0087

Tech Tips Hot Line: (800) 876-5292

Internet: http://www.techam.com

The latest TechAmerica full-color 550-page "Your Electronics Resource" catalog offers components, hardware, etc. in over 50 categories, with a 4-page small-print index. No minimum order. $4 total shipping for most parts.

APPENDIX C

Fred's Funnies

Those of us who have had the pleasure of working with Fred know that he has a big supply of funny stories to tell. Here are some we wanted to share with the readers of this book.

—the LLH Technology Publishing staff.

Confessions of an Electronic Genius

(This story is EXACTLY as written over 26 years ago for an amateur radio "ham" magazine called *73*. Little has changed since, except transistors, integrated circuits, and digital electronics have added to the mystery... and now I'm called a "computer genius"! NOT! Many of you will know what I mean....)

Have you ever been asked to fix a single-sideband transmitter, even though you weren't really sure how a simple oscillator works? Well, I have. In fact, I'm always being asked questions I shouldn't be asked. Why? Because in the minds of some around me, despite my claims to the contrary, I am an electronic genius!

How did I achieve this status? How can you attain for yourself the dubious distinction of being an "electronic genius?" Well, if you promise not to blab it around, here's the story. . . .

The Genius Is Born

I suppose it all started when I decided to build my own radio-control equipment for a model airplane. The fact that I knew nothing about electronics didn't stop me; I was surrounded at work by electronic geniuses who could solve virtually any problem involving the lowly electron—or so I thought.

Anyhow, the kit I bought was a real collection of mysterious goodies: wire, coils, tubes, phenolic, and those cute little cylindrical things with the pretty colored bands. I meticulously followed the instructions and sketches in the assembly of the receiver, a simple "single-tube super-regenerative receiver," according to the description.

Since I had no equipment to check out its operation, I took it to work for the electronic geniuses to fire-up. They performed their usual mystical rites with strange looking devices. The receiver refused to be impressed by the display. . . . and just did not work!

The next two weeks were almost too painful to describe. Complete lunch hours were consumed in discussion, theory and testing by the geniuses. My greatest contribution was keeping my fingers crossed. The geniuses, individually and collectively, all had their chance at trying to seduce "Fred's Folly" into operation. Words like superheterodyne, intermediate frequency, converter, and mixer were generously sprinkled throughout their discussions. "But," I kept repeating, "this is a super-regenerative receiver!"

The geniuses thought I had flipped. "Regenerative receivers went out with the Model T," they said, patting me on the head sympathetically.

Well, they finally gave up, and I was about to take up basket weaving as a new endeavor, when a hot spell proved fortunate. I noticed one of the silver-colored cartridge-shaped things in the receiver was leaking at one end, apparently from the heat. Could this be a bad part? It was marked ".01 MFD 100V." When this unit was replaced, the receiver worked. I had fixed it! The geniuses just shook their heads. "You are truly an electronic genius," they confided.

The Genius Grows

The bug had bitten. More receivers, more transmitters. . . . and many more problems. Somehow, never really knowing how or why, I always managed to stumble on a solution. Pretty soon I found myself fixing other guys' equipment. You've heard the expression "the blind leading the blind". . . .

About this time I decided to really find out what electronics was all about. Somehow I was not able to find anyone who was willing to sit down with me for twenty minutes and tell me all there is to know about electronics. So I attended night classes at the local high school, where I got to twirl knobs in the lab. I bought test equipment with knobs of my own to twirl. I repaired every radio the neighbors found in their attics. And, most important of all, I subscribed to *73 Magazine.*

My reputation grew. Radio repairing is, after all, mostly tube changing, dial-cord restringing, replacement of obviously cooked parts, and a generous seasoning of good luck. (Knowing what you're doing can replace the good luck; in my case, the good luck was the essential ingredient.) "You," they would tell me, "are an electronic genius!" By this time I was able to identify at least three different kinds of parts.

The Genius and the Theory

I found myself more and more becoming a victim of the never expressed, but universally accepted, theory of the masses: "He who knows anything about electronics knows everything about electronics." There is, however, a lesser known corollary to this theory: "He who knows anything about any particular branch of electronics knows practically nothing about any other branch of electronics!" I couldn't convince anyone that the latter theory more expressed my capabilities. "If it plugs into the wall, or uses a battery, Fred knows all about it," they insisted.

The Genius Takes to the Air

Then I got my ham ticket. That really did it! When my roof began sprouting weird antennas, and the neighbors' TV sets began acting in a strange manner, they were more convinced than ever that another Steinmetz was their private electronic consulting engineer. I was asked about everything from ailing TV sets (I carry service insurance on my own set) to improperly operating electric blankets (when mine quit recently, I bought a new one). And it doesn't end there; I've even found myself answering questions on the air about how to plate-modulate a transmitter, or how to eliminate chirp on CW. Sometimes I have some idea what I'm talking about, but certainly not always. However, if I tell them I don't know what I'm talking about, then I am considered overly modest; if I offer no suggestions, the conclusion is that I don't care enough to even think about the problem. A dilemma. I have found it easier to give them an answer they don't understand than to try to convince them that I'm talking through my chapeau.

The Genius Goes Stereo

Take the other night, for instance. Andy, who has known me long enough to know better, brought over a stereo tape recorder he had just built from a kit. . . . his first tussle with electronics. He said that the left channel was dead. Not being a tape recorder specialist, or any other kind of specialist, I did the only thing I could think of at the moment; I plugged in the "kluge" and turned it on. Music poured forth from both channels, loud and clear.

"What did you do to it?" Andy asked.

"Nothing," I replied.

"There you go being modest again," he said. "All you electronic geniuses are alike."

Then we tried to record. No erase. So I unbuttoned the whole works and looked at the maze of wire and stuff and things inside the chassis. I noticed two shielded cables from the erase head terminating in two plugs on the chassis. On a wild hunch (my usual method) I swapped the two plugs in their sockets. This cured the trouble. To Andy, this was sheer wizardry. When I tried to explain the four-track stereo system, and the operation of the record and erase oscillator, he absorbed about as much as a third grader trying to learn the Pythagorean Theorem.

That's about the time the left-channel playback went dead. I had no recourse but to resort to the scientific approach. Using the dirty wooden handle of a small, dirty paintbrush that happened to be lying on my dirty workbench, I pushed and shoved everything in sight under the chassis. Responding to this precision trouble-shooting technique, the left channel burst forth in full bloom. More probing disclosed that a single strand of shielding had lodged itself against the grid of the left channel pre-amp tube!

Now the left-channel magic-eye record level indicator tube was acting oddly. Andy was obviously right-handed! No amount of pushing and shoving with the paintbrush handle did any good. This exhausted my supply of magic tricks, so I suggested that we put the whole works back in the case and be glad that it hadn't gone up in smoke. All buttoned up, we gave it the final check. No one was more surprised than I when everything worked, including the left-channel magic-eye indicator!

"You did something when I wasn't looking," accused Andy.

With a knowing expression, I replied, "The hand is quicker than the 'eye,' my friend."

"K.I.S.S. or K.I.C.K.?"

Many times in organizing a task, such as setting up a data file, designing a computer program, selecting new equipment, or managing your hard disk, you are faced with decisions that can either simplify or complicate the task. Do you K.I.S.S. it or K.I.C.K. it?

You have probably heard of the K.I.S.S. approach—"Keep It Simple, Stupid!" You may give this a lot of lip service, while you dream up bigger and better complexities. Instead of KISSing the job, you are probably KICKing it. K.I.C.K.? That stands for "Keep It Confusing, Knucklehead!"

After all, if you confuse everyone around you, it protects your job. They'll think you're so smart they can't get along without you. You become the local "guru" by being a KICKer rather than a KISSer.

I've been amazed over the years to see so many examples of KICKers in action, even outside the computer world. Most doctors, lawyers, accountants and engineers are KICKers rather than KISSers. Psychologists and psychiatrists are big KICKers. Probably the biggest KICKers of all are government workers and bureaucrats at all levels. How about politicians, who, after all, usually got their KICK training as lawyers?

In the musical world, in 1934 Cole Porter wrote "I Get a KICK Out of You," with "flying too high" as part of the lyrics. Apparently he changed his thinking by 1948 when he wrote the Broadway musical "KISS Me, Kate."

I was aware of the deep entrenchment of KICKers in the corporate structure when I worked in the aerospace industry as an engineer, surrounded by KICKers at all levels. Every time I tried to KISS, I'd get KICKed!

Cost-plus-fixed-fee contracts don't benefit from KISSing, only KICKing.

However, when there is an incentive to KISS, complexity gets KICKed out. Businessmen — especially self-employed entrepreneurs — are usually KISSers. Why? Because they are dealing with their own bucks! The more they KISS, the bigger the "bottom line."

KISSing the KICKers

I taught myself electronics, many years ago, by reading *Popular Electronics, Radio-Electronics, Electronics Illustrated,* and other electronic magazines (most of them gone today.) I was frequently appalled at the complexity of some projects that could be done with two or three transistors instead of the dozen used in some KICKer articles. So I started writing KISS articles. I became known for my KISSing.

Then I got into my own part-time Amway distributor networking business. I started KICKing until I saw how much time and effort I was wasting on needless tasks. As I saw negative cash flow, I became the biggest KISSer in town! The business became so successful, I retired from aerospace after five years—and haven't worked for anyone else in the 23 years since.

When the paperwork in our growing business got overwhelming, I bought a small computer and looked for programs for my wife to use. She's a great KISSer! But all the programmers I found were KICKers. They wanted to address the power of the computer instead of the simplicity of our needs. I KISSed them off and wrote my own programs.

My programs were simple to learn and simple to use, a KISSers delight. Originally written in 1980 for a 16K RAM TRS-80 Model I, my eight programs address the needs of the business rather than the capabilities of a computer.

How to KISS

I don't keep the prices of thousands of product prices in memory. That's too slow and awkward, so I KISS the top 400, and the program easily handles the others. I don't do inventory control; it's easier to look on the shelf. I don't keep track of each distributor's volume of business all month. It's faster to add the order totals on a calculator for the relative few that qualify for bonuses based on volume. I don't do double-entry bookkeeping or generate a profit and loss statement; I just keep putting money in the bank. The names and addresses of my distributors are still on 3x5 file cards. I don't know (and don't care) how many bottles or boxes of particular products I've sold in the last three years, by customer, by date, by amount, or by any other parameter. I'm a KISSer, and have better things to do than analyze things that don't matter. If I can't KISS it, I don't do it.

The "proof of the pudding" is that my wife, Ev, who HATES computers, uses these programs for distributor order processing and bonus statements, customer invoices, and bookkeeping—and now has TWO computers! She tells me that if it weren't for the computers and my simple programs, I would be doing the paperwork!

Wanna' Become a LOVER?

This whole KISS philosophy will probably shock some of you—especially those of you who use a $400 computer program with a Pentium Pro computer, 2.5 gigabyte hard disk, and 64 megs of RAM to balance your check book. You may be using a powerful database program to keep track of your Christmas Card List or recipes, or a $500 word processing program and $2000 laser printer for your monthly letter to your parents. Would you buy a big truck to carry a bottle of milk from the store? Is it possible that's what you're doing with your computer?

Next time you're faced with a task, ask yourself a few questions. Is this the simplest way to do this job? Is this "overkill"? Isn't there an easier way? Would 3x5 cards be simpler? Would you rather be a KISSer or a KICKer? Remember, KISSers eventually become LOVERs ("Leave Out Various Extras, Rube!").

Memories Are Made of This

I don't know about you, but I'm confused about the various kinds of computer memory. Someone, somewhere seems to get some kind of thrill by creating names for memory (and most other computer-related hardware) that defy explanation. Furthermore, the names sound so much alike that more confusion results.

We've all heard of "RAM" (random access memory) and "ROM" (read-only memory.) Sound alike, don't they? They even LOOK alike, integrated circuit chips that can only be told apart by their mysterious markings.

How about "expanded memory" and "extended memory"? After reading at least a dozen articles about these, I've decided that even the article authors don't really know the similarities and differences between these, especially when software can apparently make one simulate the other!

So I've decided to add to the confusion with some new memory nomenclature. See if you can keep these in YOUR memory.

Expended Memory: This is the computer's memory you are already using. Some is RAM, some is ROM, some is expanded, and some is extended.

Distended Memory: This is expended expanded extended memory beyond the address capabilities of the computer unless you add math and graphics coprocessors.

Unattended Memory: This is memory you're not yet using, but available in your machine. Everything beyond expended expanded extended distended memory. May be used to run flight simulators in background mode while your boss is nearby.

Intended Memory: This is what you'll need to run Windows98 and all future applications from Microsoft. You thought you could survive with 64 megs? Don't be ridiculous!

Pretended Memory: This is what you thought you had when you bought your computer and later found out would require adding some expensive chips. Typically, this is having a sixteen megabyte memory board in a machine with only one megabyte of RAM actually installed.

Tremendous Memory: Many gigabytes.

Stupendous Memory: Beyond tremendous.

Defended Memory: This has something to do with the 80386 "protected mode." (Don't ask me; I don't even work here.)

Contented Memory: This is the portion of memory where all your happy little TSR programs joyfully reside, snugly hidden away from your applications, but ready to jump into action to please you—or interfere with the program you're trying to run!

Frequented Memory: Your disk cache stays here, often used and constantly on call.

Demented Memory: The part of memory reserved for thriller games like *Dungeons and Dragons.*

Suspended Memory: This is where your program is hiding when your computer "goes out to lunch."

Surrendered Memory: This is the memory you recover when you hit the RESET button.

Out-Of Memory: This is the no-memory land to which most of my programs seem to find their way.

And here are some more memory names for Microsoft to use, as yet undefined: enchanted, lamented, amended, offended, blended, fomented, and rescinded. Do you have some others?

Fred's Twenty-Five Mistresses

by Ev Blechman

(As you can tell, this next story was written by Fred's wife, Ev!)

My husband, Fred, has had 25 mistresses. At least, that's all I know about. Of course, they say the wife is the last to know, but not in my case.

It happened! Not one, not two, not three, not four, but 25 times! And the sad thing is that I found out about the first one the moment he carried HER across our threshold.

It was a cold January morning in 1978. Fred and I had only been married two months when he brought HER right to our doorstep. She was well disguised, but I was immediately intimidated by her because she weighed less than I did!

Fred carried her over the threshold with no explanation and then, just like the ravenous beauty that pops out of the cake at a bachelor's last gig, SHE popped out of her carefully designed outer garments.

Physically, my contours were much more curvy than hers. She was small, definitely angular, but solid. I should have deduced that from her outer garments. While I try to maintain a good California tan and a sunny disposition, she appeared in chic gray with black and silver accessories, and had a distinctive cool and precise manner.

The delicate way in which Fred handled—and fondled—her should have warned me of the constant confrontations which were to become part of our future, and which I've endured now for 15 years. I admit to a slight scowl as I noticed Fred handling her so gingerly, and carefully, like a thing of great fragility. I didn't take his pulse, but it must have been racing as he picked her up, turned her over, inspected her closely, and gently ran his hands all over her frame.

She arrived, reminiscent of a bride and her trousseau, with various accessories. Fred immediately took charge, attaching the accessories in the right places. I couldn't help taking note of his excited manner, his joy, the wondrous expression of anticipation of glorious hours that he would spend with her. . . . and I was jealous.

Because I was Fred's new bride, there was a strong obligation on Fred's time; I expected it to be spent with me! When we got married he said "Stick with me, Babe. We'll get rich building our Amway business together."

I won out for a short while. Then came the long, lonely hours waiting for him to leave HER side. I never knew when I would see the light go out in the guest room where he had made a home for her. Many times, late into the night, I would reach out to touch him, only to find that he was not there. However, he was always at my side when I woke up in the morning—but dead to the world!

I was beside myself. I had overheard Fred tell a friend on the telephone that he was teaching her to do just about anything he wanted. No wonder he appeared so haggard as he dragged himself into bed in the wee hours of the morning.

I quickly discovered a marked distinction between us. I didn't have all the answers, she did. And my Fred was trying to find them! I built up quite a resentment over this third party who was spending twenty-four hours a day in our new-bride paradise.

Something had to be done. I'd had it. Fred was being "wenched" away from me. One night I stomped my way towards the guest room and flung open the door. Just as I suspected, there was Fred, hovering over her as he had day after day for the last six months. She remained still—not a move. But I heard her humming to him. My presence was completely ignored. I hurled out of the room, slammed the door behind me and waited for some response. None. They were totally engrossed doing their thing together.

Hours later Fred came into our bedroom and announced he'd done all he could with her, and that he'd need to get another more powerful "model." I said nothing as my fury built up. He had made up his mind and my silence was mistaken for acquiescence. He got three more. All from the same family!

As time went by, Fred had 25 of these—you guessed it—microcomputers under our roof (although, thank goodness, not all at the same time!). Instead of being restricted to just the guest room, Fred used a total of three bedrooms, and spent more and more time running from one bedroom to the other. Over a period of time, some left as new ones arrived.

Just imagine how you would feel if you saw your home was invaded first by a TRS-80 Model I, then three Model IIIs, two Model 4s, two Model 4Ps, two Sinclair ZX-81s, a Timex Sinclair 1500, a Timex Sinclair 2068, a Radio Shack MC10, a Coleco ADAM, an Apple IIc, a Sanyo MBC 555, a Sinclair QL, two Sinclair Spectrums, an IBM PC/XT Clone, a 286 clone, a book size PC/XT, a Toshiba 1000 laptop, a Laser PC4 notebook, and a Microgold 286 portable. (Know any wife who would learn the names of her husband's 25 mistresses?)

At this point in time, there are only 12 micros left, and like a family, they share some of the bedrooms, the play room, and the office. It took a while to convince me that I would be able to accept and eventually love, if not all, one or two of Fred's "mistresses." Believe it or not, I can't wait to get my hands on one every day when I do paperwork for our Amway business with the $39 "AMBIZ-PAK" of eight programs Fred wrote while I thought he was "fooling around." Maybe a wife IS the last one to know.......

About the Author: Ev Blechman is a former professional dancer and movie script writer. She and Fred are retired Amway Emerald Direct Distributors.

Blechman's Ten Laws of Computing

I've owned 26 microcomputers, I've written over 500 magazine articles and five books specifically about microcomputer hardware and software since 1978, and I've come to the following conclusions, which I call "Blechman's Ten Laws of Computing":

1) "When it's manufactured, it's already 'obsolete'—but still far more powerful than you need."
2) "When you try to use it, it's incompatible with everything you have that used to work."
3) "When you try to return it, they're out of business—or suddenly don't understand English."
4) "No matter how big your hard drive, it will be filled within 30 days or less, mostly with things you'll never use."
5) "Nothing works the first time, and never works when you try to show it off."

6) “Everything you use is attacking your body with electromagnetic radiation of various sorts.”

7) “Any software upgrade costing less than $20 is an admission of guilt.”

8) “Version 1 of any software is full of ‘bugs.’ Version 2 fixes all the bugs and is great. Version 3 adds all the things users ask for, but hides all the great stuff in Version 2.”

9) “Any software costing over $100, or with documentation of over 100 pages, is too complicated.”

10) “You probably were better off with 3x5 cards and a typewriter in the first place!”

There are ways to beat these “laws,” but that could be the subject of a l-o-n-g article (and a one-hour talk I've given to computer local clubs).

I only offer these “laws” here to counter the enormous computer hype that is avalanching unsuspecting buyers who might well be satisfied with plain-vanilla used PCs, XTs or 286 machines with monochrome monitors.

Word Certain 2.0—The Ultimate Word Processor?

My friend, Mel Marcus, is a software freak! When he had his Commodore 64 years ago he collected hundreds of public domain and shareware programs. Then, when he got into the IBM PC world, he went nuts with shareware programs from computer bulletin board systems (BBSs) for years, downloading everything he could find.

Lately, like a growing number of cyber-freaks, Mel spends hours and hours roving around the electronic black hole known as the Internet, with its thousands of BBSs and “newsgroups” and their hundreds of thousands of downloadable files. So when Mel tells me he has found something new and exciting, I listen.

Mel called me the other day and said “Fred, I've finally found it! I found it!” He got my attention.

“What did you find, Mel? I didn't know you were looking for anything.”

“Oh, yes!” he explained. “Ever since I started using microcomputers, I've been looking for a great word processor. I've found some good ones, but they were always too complicated. I do a lot of writing in my business, and I need something fast and simple.

“With the Commodore I used Paper Clip,” he continued. “But since getting my first PC, I've found so many word processors to choose from, it's bewildering. Of course, there's WordStar and all its workalikes. I've tried MultiMate, VolksWriter, Leading Edge, WordVision, LeScript, The Word, WordPerfect, Word for Windows, and just about every one around. In the shareware world, I found PC Write to be about the most powerful.”

“But,” he continued, “they are all either too limited or too complicated. The manuals are hundreds of pages long. Printer installations can drive you bananas. And just about the time you learn how to use one of these word processors, they put out a new version that won't read the old text files! And the Windows versions—well, they confuse me altogether. So many buttons, bars, icons, and dialogue boxes, I get totally lost and can't find the things I used to use in DOS."

"I know what you mean," I replied. "But you have finally found one you like? Tell me about it."

"Well," Mel started out, "it's called Word Certain 2.0. I've never seen it advertised, but I ran across it on the Internet. Best of all, it's not even shareware—it's public domain, so you don't even have to pay anyone to use it! It was written by a guy named Kevin Mitnick. I've got it on disk. You can come over and run the disk yourself. It's fabulous! I even wrote out a list of features that. . . ."

"Hold it," I interrupted. "Can you send me that list right away? The deadline for the next issue of *Nuts & Volts* is coming up. Maybe I can tell the world about Word Certain 2.0."

"Sure, Fred. I'll drop it in the mail tonight."

A couple of days later I received the list from Mel. It's shown here:

WORD CERTAIN 2.0 Main Features

1. Instant access. No waiting for DOS to load.
2. All commands internal. No accessing disk for commands or overlay programs. No extended memory required.
3. 270K storage capacity (about 45,000 words, or 180 double-spaced pages.)
4. Each character immediately stored to reduce chance of data loss.
5. Instant error correction with included utility.
6. Full search and replace functions.
7. Thesaurus and dictionary compatible.
8. Desktop publishing ability lets user specify layout, and freely create and import graphics. Automatic hand-scanning and image editing.
9. Prints out in selected colors.
10. Simple, easily understood commands.
11. Produces upper and lower case English and foreign characters, as well as proof-reading symbols.
12. Prints in many different fonts and sizes. Especially efficient for script fonts.
13. Does not require special printer installation or paper feed mechanism. Works with pin-feed or cut sheet paper.
14. Subscripts, superscripts, underlining, bold, enhanced.
15. Insert, delete, indents and other formatting functions.
16. Cut and paste for block moves.
17. Page numbering, headers, and footers.
18. Automatic on-line real-time spelling and grammar checker.
19. Import from Lotus, Dbase and other popular programs.
20. Can be modified by user without learning new language.

21. Easily transportable between different computers.
22. Does not conflict with memory-resident programs.
23. Multitasking. Can be used while computer is running a background program.
24. Data easily duplicated and shared with other users.
25. Can also create spreadsheets, forms, and data bases with no additional software.
26. Works with Mac or PC without modification.

I was impressed. I called Mel. "Mel, you say you have Word Certain on disk. Can you send me a copy? I'd like to see for myself. I find it hard to believe. Can it really do all those things on your list?"

"Hey, Fred, it's unbelievable! Everyone has been after me for copies and it's been driving me crazy. I don't make a nickel on this. I'm out of blank disks and mailers. Why don't you come over and I'll let you run the disk?"

My curiosity was at a peak. This sounds like a natural article for *Nuts & Volts.* Larry and Robin love new stuff! Mel's place is only a few miles from me, so I made an appointment with him that afternoon.

When I got to Mel's door I recognized the yellow pad and pencil he always had on his door for notes. There was a note for me: "Fred: Had to leave for a few minutes. Come on in. The computer is on and ready."

I went inside and there was Mel's 486 multimedia monstrosity. On the screen it said, "Type DIR, press Enter, then look at README.1ST." I typed DIR and pressed the Enter key. The screen showed a bunch of files like WC.EXE, WCINSTAL.EXE, WCHELP.DOC and README.1ST. I typed TYPE README.1ST and pressed Enter. The screen went blank then displayed this message:

LOOK AT THE CALENDAR ON THE WALL, THEN TURN AROUND.

I looked at the calendar. It said April 1. I turned around to see Mel standing there, grinning from ear to ear, and holding out a yellow pad and a slim, yellow pencil with a nice rubber eraser on top.

"Here's Word Certain 2.0. Like I told you, it's unbelievable. See," he said, holding out the pencil, "the top is software, the wooden part is hardware, and the point is the printer. The world's simplest word processor."

He's probably right! How can you beat a pencil and paper for simplicity?

Index

Notes